An Introduction to
Multivariable Mathematics

An Introduction to Multivariable Mathematics

Leon Simon

www.morganclaypool.com

ISBN: 9781598298017   paperback
ISBN: 9781598298024   ebook

DOI 10.2200/S00147ED1V01Y200808MAS003

A Publication in the Morgan & Claypool Publishers series
SYNTHESIS LECTURES ON MATHEMATICS AND STATISTICS

Lecture #3
Series Editor: Steven G. Krantz, Washington University, St. Louis

Series ISSN
Synthesis Lectures on Mathematics and Statistics
ISSN pending.

An Introduction to Multivariable Mathematics

Leon Simon

Stanford University

SYNTHESIS LECTURES ON MATHEMATICS AND STATISTICS #3

MORGAN & CLAYPOOL PUBLISHERS

ABSTRACT

The text is designed for use in a 40 lecture introductory course covering linear algebra, multivariable differential calculus, and an introduction to real analysis.

The core material of the book is arranged to allow for the main introductory material on linear algebra, including basic vector space theory in Euclidean space and the initial theory of matrices and linear systems, to be covered in the first 10 or 11 lectures, followed by a similar number of lectures on basic multivariable analysis, including first theorems on differentiable functions on domains in Euclidean space and a brief introduction to submanifolds. The book then concludes with further essential linear algebra, including the theory of determinants, eigenvalues, and the spectral theorem for real symmetric matrices, and further multivariable analysis, including the contraction mapping principle and the inverse and implicit function theorems. There is also an appendix which provides a 9 lecture introduction to real analysis. There are various ways in which the additional material in the appendix could be integrated into a course—for example in the Stanford Mathematics honors program, run as a 4 lecture per week program in the Autumn Quarter each year, the first 6 lectures of the 9 lecture appendix are presented at the rate of one lecture a week in weeks 2–7 of the quarter, with the remaining 3 lectures per week during those weeks being devoted to the main chapters of the text.

It is hoped that the text would be suitable for a 1 quarter or 1 semester course for students who have scored well in the BC Calculus advanced placement examination (or equivalent), particularly those who are considering a possible major in mathematics.

The author has attempted to make the presentation rigorous and complete, with the clarity and simplicity needed to make it accessible to an appropriately large group of students.

KEYWORDS

vector, dimension, dot product, linearly dependent, linearly independent, subspace, Gaussian elimination, basis, matrix, transpose matrix, rank, row rank, column rank, nullity, null space, column space, orthogonal, orthogonal complement, orthogonal projection, echelon form, linear system, homogeneous linear system, inhomogeneous linear system, open set, closed set, limit, limit point, differentiability, continuity, directional derivative, partial derivative, gradient, isolated point, chain rule, critical point, second derivative test, curve, tangent vector, submanifold, tangent space, tangential gradient, permutation, determinant, inverse matrix, adjoint matrix, orthonormal basis, Gram-Schmidt orthogonalization, linear transformation, eigenvalue, eigenvector, spectral theorem, contraction mapping, inverse function, implicit function, supremum, infimum, sequence, convergent, series, power series, base point, Taylor series, complex series, vector space, inner product, orthonormal sequence, Fourier series, trigonometric Fourier series.

Contents

Preface

The present text is designed for use in an introductory honors course. It is in fact a fairly faithful representation of the material covered in the course "Math 51H—Honors Multivariable Mathematics" taught during the 10 week Autumn quarter in the Stanford Mathematics Department, and typically taken in the first year by those students who have scored well in the BC Calculus advanced placement examination (or equivalent), many of whom are considering a possible major in mathematics.

The text is designed to be covered by 4 lectures per week (each of 50 minutes duration) in a 10 week quarter, or by 3 lectures per week over a 13 week semester. In the Stanford program the material in the Appendix ("Introductory lectures on real analysis") is run parallel to the material of the main text—the first 6 real analysis lectures are presented in doses of 1 lecture per week (Mondays are "real analysis days" in the Stanford program) over a 6-week period, and during these weeks the remaining 3 lectures per week are devoted to the main text. A typical timetable based on this approach is to be found at:

http://math.stanford.edu/~lms/51h-timetable.htm

A serious effort has been made to ensure that the present text is rigorous and complete, but at the same time that it has the clarity and simplicity needed to make it accessible to an appropriately large group of students. For most students, regular (at least weekly) problem sets will be of paramount importance in coming to grips with the material, and ideally these problem sets should involve a mixture of questions which are closely integrated with the lectures and which are designed to help in the assimilation the various concepts at a level which is at least one step removed from routine application of standard definitions and theorems.

This text is dedicated to the many Stanford students who have taken the Honors Multivariable Mathematics course in recent years. Their vitality and enthusiasm have been an inspiration.

Leon Simon
Stanford University
July 2008

CHAPTER 1

Linear Algebra

1 VECTORS IN $\mathbb{R}^n$

$\mathbb{R}^n$ will denote real Euclidean space of dimension n—that is, $\mathbb{R}^n$ is the set of all ordered n-tuples of real numbers, referred to as "points in $\mathbb{R}^n$" or "vectors in $\mathbb{R}^n$"—the two alternative terminologies "points" and "vectors" will be used quite interchangeably, and we in no way distinguish between them.[1] Depending on the context we may decide to write vectors (i.e., points) in $\mathbb{R}^n$ as columns or rows: in the latter case, $\mathbb{R}^n$ would denote the set of all ordered n-tuples $(x_1, \ldots, x_n)$ with $x_j \in \mathbb{R}$ for each $j = 1, \ldots, n$. In the former case, $\mathbb{R}^n$ would denote the set of all columns $\begin{pmatrix} x_1 \\ x_2 \\ \vdots \\ x_n \end{pmatrix}$. In this case we often (for compactness of notation) write $(x_1, \ldots, x_n)^{\mathrm{T}}$ as an alternative notation for the column vector $\begin{pmatrix} x_1 \\ x_2 \\ \vdots \\ x_n \end{pmatrix}$, and conversely $\begin{pmatrix} x_1 \\ x_2 \\ \vdots \\ x_n \end{pmatrix}^{\mathrm{T}} = (x_1, \ldots, x_n)$.

For the moment, we shall write our vectors (i.e., points) as rows. Thus, for the moment, points $\underline{x} \in \mathbb{R}^n$ will be written $\underline{x} = (x_1, \ldots, x_n)$. The real numbers $x_1, \ldots, x_n$ are referred to as *components* of the vector $\underline{x}$. More specifically, x_j is *the j-th component of $\underline{x}$*.

To start with, there are two important operations involving vectors in $\mathbb{R}^n$: we can (i) add them according to the definition

1.1 $$(x_1, \ldots, x_n) + (y_1, \ldots, y_n) = (x_1 + y_1, \ldots, x_n + y_n)$$

and (ii) multiply a vector $(x_1, \ldots, x_n)$ in $\mathbb{R}^n$ by a scalar (i.e., by a $\lambda \in \mathbb{R}$) according to the definition

1.2 $$\lambda(x_1, \ldots, x_n) = (\lambda x_1, \ldots, \lambda x_n).$$

Using these two definitions we can prove the following more general property: If $\underline{x} = (x_1, \ldots, x_n)$, $\underline{y} = (y_1, \ldots, y_n)$, and if $\lambda, \mu \in \mathbb{R}$, then

1.3 $$\lambda \underline{x} + \mu \underline{y} = (\lambda x_1 + \mu y_1, \ldots, \lambda x_n + \mu y_n).$$

[1]Nevertheless, from the intuitive point of view it is sometimes helpful to think geometrically of vectors as directed arrows; for example, such a geometric interpretation is suggested in the discussion of *line vectors* below.

Notice, in particular, that, using the definitions 1.1, 1.2, starting with any vector $\underline{x} = (x_1, \ldots, x_n)$ we can write

1.4
$$\underline{x} = \sum_{j=1}^{n} x_j \underline{e}_j \, ,$$

where $\underline{e}_j$ is the vector with each component $= 0$ except for the j-th component which is equal to 1. Thus, $\underline{e}_1 = (1, 0, \ldots, 0), \underline{e}_2 = (0, 1, 0, \ldots, 0), \ldots, \underline{e}_n = (0, \ldots, 0, 1)$. The vectors $\underline{e}_1, \ldots, \underline{e}_n$ are called *standard basis vectors for* $\mathbb{R}^n$.

The definitions 1.1, 1.2 are of course geometrically motivated—the definition of addition is motivated by the desire to ensure that the triangle (or parallelogram) law of addition holds (we'll discuss this below after we introduce the notion of "line vector" joining two points of $\mathbb{R}^n$) and the second is motivated by the natural desire to ensure that $\lambda \underline{x}$ should be a vector in either the same or reverse direction as $\underline{x}$ (according as λ is positive or negative) and that it should have "length" equal to $|\lambda|$ times the original length of $\underline{x}$. To make sense of this latter requirement, we need to first define length of a vector $\underline{x} = (x_1, \ldots, x_n) \in \mathbb{R}^n$: motivated by[2] Pythagoras, it is natural to define the length of $\underline{x}$ (also referred to as *the distance of $\underline{x}$ from* $\underline{0}$), denoted $\|\underline{x}\|$, by

1.5
$$\|\underline{x}\| = \sqrt{\sum_{j=1}^{n} x_j^2} \, .$$

There is also a notion of "line vector" $\overrightarrow{AB}$ from one point $A = \underline{a} = (a_1, \ldots, a_n) \in \mathbb{R}^n$ to another point $B = \underline{b} = (b_1, \ldots, b_n)$ defined by

1.6 $\qquad \overrightarrow{AB} = (b_1 - a_1, \ldots, b_n - a_n) \quad (= \underline{b} - \underline{a}$ using the definition 1.3) .

Notice, that geometrically, we may think of $\overrightarrow{AB}$ as a directed line segment (or arrow) with tail at A and head at B, but mathematically, $\overrightarrow{AB}$ is nothing but $\underline{b} - \underline{a}$, i.e., the point with coordinates $(b_1 - a_1, \ldots, b_n - a_n)$. By definition, we then do have the triangle identity for addition:

1.7 $\qquad\qquad\qquad \overrightarrow{AB} + \overrightarrow{BC} = \overrightarrow{AC} \, ,$

because if $A = \underline{a} = (a_1, \ldots, a_n), B = \underline{b} = (b_1, \ldots, b_n), C = \underline{c} = (c_1, \ldots, c_n)$ then $\overrightarrow{AB} + \overrightarrow{BC} = (\underline{b} - \underline{a}) + (\underline{c} - \underline{b}) = \underline{c} - \underline{a} = \overrightarrow{AC}$.

---

[2]There is often confusion here in elementary texts, which are apt to claim, at least in the case $n = 2, 3$, that 1.5 is a consequence of Pythagoras, i.e., that 1.5 is proved by using Pythagoras' theorem; that is not the case—1.5 is *a definition, motivated by the desire to ensure that the Pythagorean theorem holds* when we later give the appropriate definition of angle between vectors.

2   DOT PRODUCT AND ANGLE BETWEEN VECTORS IN $\mathbb{R}^n$

If $\underline{x} = (x_1, \ldots, x_n)$ and $\underline{y} = (y_1, \ldots, y_n)$, we define the dot product of $\underline{x}$ and $\underline{y}$, denoted by $\underline{x} \cdot \underline{y}$, by $\underline{x} \cdot \underline{y} = \sum_{j=1}^{n} x_j y_j$. Notice that then the dot product has the properties:

(i)
$$\underline{x} \cdot \underline{y} = \underline{y} \cdot \underline{x}$$

(ii)
$$(\mu \underline{x} + \lambda \underline{y}) \cdot \underline{z} = \mu \underline{x} \cdot \underline{z} + \lambda \underline{y} \cdot \underline{z}$$

(iii)
$$\underline{x} \cdot \underline{x} = \|\underline{x}\|^2 \, .$$

Using the above properties, we can check the identity

2.1
$$\|\underline{x} + \underline{y}\|^2 = \|\underline{x}\|^2 + \|\underline{y}\|^2 + 2\underline{x} \cdot \underline{y}$$

as follows:

$$
\begin{aligned}
\|\underline{x} + \underline{y}\|^2 &= (\underline{x} + \underline{y}) \cdot (\underline{x} + \underline{y}) && \text{(by (iii))}\\
&= \underline{x} \cdot \underline{x} + \underline{y} \cdot \underline{y} + 2\underline{x} \cdot \underline{y} && \text{(by several applications of (i) and (ii))}\\
&= \|\underline{x}\|^2 + \|\underline{y}\|^2 + 2\underline{x} \cdot \underline{y} && \text{(by (iii) again)} \, .
\end{aligned}
$$

In particular, with $t\underline{y}$ in place of $\underline{y}$ (where $t \in \mathbb{R}$) we see that

2.2
$$\|\underline{x} + t\underline{y}\|^2 = t^2 \|\underline{y}\|^2 + 2t\underline{x} \cdot \underline{y} + \|\underline{x}\|^2 \, .$$

Observe that if $\underline{y} \neq \underline{0}$ this is a quadratic in the variable t, and by using "completion of the square" $at^2 + 2bt + c = a(t + b/a)^2 + (ac - b^2)/a$ (valid if $a \neq 0$), we see that 2.2 can be written as:

2.3
$$\|\underline{x} + t\underline{y}\|^2 = \|\underline{y}\|^2 (t + \underline{x} \cdot \underline{y}/\|\underline{y}\|)^2 + (\|\underline{y}\|^2\|\underline{x}\|^2 - (\underline{x} \cdot \underline{y})^2)/\|\underline{y}\| \, .$$

Observe that the right side here evidently takes its minimum value when, and only when, $t = -\underline{x} \cdot \underline{y}/\|\underline{y}\|$ and the corresponding minimum value of $\|\underline{x} + t\underline{y}\|^2$ is $(\|\underline{y}\|^2\|\underline{x}\|^2 - (\underline{x} \cdot \underline{y})^2)/\|\underline{y}\|$ and the corresponding value of $\underline{x} + t\underline{y}$ is the point $\underline{x} - \|\underline{y}\|^{-2}(\underline{x} \cdot \underline{y})\underline{y}$. Thus,

2.4
$$\|\underline{x} + t\underline{y}\|^2 \geq (\|\underline{y}\|^2\|\underline{x}\|^2 - (\underline{x} \cdot \underline{y})^2)/\|\underline{y}\| \, ,$$

with equality if and only if $t = -\underline{x} \cdot \underline{y}/\|\underline{y}\|$, in which case $\underline{x} + t\underline{y} = \underline{x} - \|\underline{y}\|^{-2}(\underline{x} \cdot \underline{y})\underline{y}$. We can give a geometric interpretation of 2.4 as follows.

For $\underline{y} \neq \underline{0}$, the straight line ℓ through $\underline{x}$ parallel to $\underline{y}$ is by definition the set of points $\{\underline{x} + t\underline{y} : t \in \mathbb{R}\}$, so 2.4 says geometrically that the point $\underline{x} - \|\underline{y}\|^{-2}(\underline{x} \cdot \underline{y})\underline{y}$ is the point of ℓ which has least distance to the origin, and the least distance is equal to $\sqrt{(\|\underline{y}\|^2\|\underline{x}\|^2 - (\underline{x} \cdot \underline{y})^2)/\|\underline{y}\|}$.

Observe that an important particular consequence of the above discussion is that $\|\underline{y}\|^2\|\underline{x}\|^2 - (\underline{x} \cdot \underline{y})^2 \geq 0$, i.e.,

2.5
$$|\underline{x} \cdot \underline{y}| \leq \|\underline{x}\| \, \|\underline{y}\| \quad \forall \, \underline{x}, \underline{y} \in \mathbb{R}^n \, ,$$

and in the case $y \neq 0$ equality holds in 2.5 if and only if $\underline{x} = \|\underline{y}\|^{-2}(\underline{x} \cdot \underline{y})\underline{y}$. (We technically proved 2.5 only in case $\underline{y} \neq \underline{0}$, but observe that 2.5 is also true, and equality holds, when $\underline{y} = \underline{0}$.) The inequality 2.5 is known as the Cauchy-Schwarz inequality.

An important corollary of 2.5 is the following "triangle inequality" for vectors in $\mathbb{R}^n$:

2.6
$$\underline{x}, \underline{y} \in \mathbb{R}^n \Rightarrow \|\underline{x} + \underline{y}\| \leq \|\underline{x}\| + \|\underline{y}\| .$$

This is proved as follows. By 2.1 and 2.5, $\|\underline{x} + \underline{y}\|^2 = \|\underline{x}\|^2 + \|\underline{y}\|^2 + 2\underline{x} \cdot \underline{y} \leq \|\underline{x}\|^2 + \|\underline{y}\|^2 + 2\|\underline{x}\|\|\underline{y}\| = (\|\underline{x}\| + \|\underline{y}\|)^2$ and then 2.6 follows by taking square roots.

Now we want to give the definition of the angle between two nonzero vectors $\underline{x}$, $\underline{y}$. As a preliminary we recall that the function $\cos\theta$ is a 2π-periodic function which has maximum value 1, attained at $\theta = k\pi$, k any even integer, and minimum value -1 attained at $\theta = k\pi$ with k any odd integer. Also $\cos\theta$ is strictly decreasing from 1 to -1 as θ varies between 0 and π. Thus, for each value $t \in [-1, 1]$ there is a unique value of $\theta \in [0, \pi]$ (denoted $\arccos t$) with $\cos\theta = t$. We will give a proper definition of the function $\cos\theta$ and discuss these properties in detail later (in Lecture 6 of Appendix A), but for the moment we just assume them without further discussion. We are now ready to give the formal definition of the angle between two nonzero vectors. So suppose $\underline{x}, \underline{y} \in \mathbb{R}^n \setminus \{0\}$. Then we define[3] the angle θ between $\underline{x}, \underline{y}$ by

2.7
$$\theta = \arccos\left(\|x\|^{-1}\underline{x} \cdot \|y\|^{-1}\underline{y}\right) .$$

Notice this makes sense because $\|x\|^{-1}\underline{x} \cdot \|y\|^{-1}\underline{y} \in [-1, 1]$ by the Cauchy-Schwarz inequality 2.5, which says precisely that $-\|\underline{x}\|\|\underline{y}\| \leq \underline{x} \cdot \underline{y} \leq \|\underline{x}\|\|\underline{y}\|$.

Observe that by definition we then have

2.8
$$\underline{x} \cdot \underline{y} = \|\underline{x}\|\|\underline{y}\| \cos\theta, \quad \underline{x}, \underline{y} \in \mathbb{R}^n \setminus \{0\} ,$$

where θ is the angle between $\underline{x}, \underline{y}$.

2.9 Definition: Two vectors $\underline{x}, \underline{y}$ are said to be orthogonal if $\underline{x} \cdot \underline{y} = 0$.

Observe that then the zero vector is orthogonal to every other vector, and, by 2.8, two nonzero vectors $\underline{x}, \underline{y} \in \mathbb{R}^n$ are orthogonal if and only if the angle between them is $\pi/2$ (because $\cos\theta = 0$ with $\theta \in [0, \pi]$ if and only if $\theta = \pi/2$).

SECTION 2 EXERCISES

2.1 Explain why the following "proof" of the Cauchy-Schwarz inequality $|\underline{x} \cdot \underline{y}| \leq \|\underline{x}\|\,\|\underline{y}\|$, where $\underline{x}, \underline{y} \in \mathbb{R}^n$, is not valid:

---

[3] Again, there is often confusion in elementary texts about what is a definition and what is a theorem. In the present treatment, 2.7 is a *definition* which relies only on the Cauchy-Schwarz inequality 2.5 (which ensures $\|x\|^{-1}\|y\|^{-1}\underline{x} \cdot \underline{y} \in [-1, 1]$) and the fact that $\arccos$ is a well defined function on $[-1, 1]$ (with values $\in [0, \pi]$). With this approach it becomes a theorem rather than a definition that $\cos\theta$ is the ratio (length of the edge adjacent to the angle θ)/(length of hypotenuse) in a right triangle—see problem 6.5 of real analysis Lecture 6 in the Appendix for further discussion.

Proof: If either $\underline{x}$ or $\underline{y}$ is zero, then the inequality $|\underline{x} \cdot \underline{y}| \le \|\underline{x}\| \, \|\underline{y}\|$ is trivially correct because both sides are zero. If neither $\underline{x}$ nor $\underline{y}$ is zero, then as proved above $\underline{x} \cdot \underline{y} = \|\underline{x}\| \, \|\underline{y}\| \cos\theta$, where θ is the angle between $\underline{x}$ and $\underline{y}$. Hence, $|\underline{x} \cdot \underline{y}| = \|\underline{x}\| \, \|\underline{y}\| \, |\cos\theta| \le \|\underline{x}\| \, \|\underline{y}\|$.

Note: We shall give a detailed discussion of the function $\cos x$ later (in Lecture 6 of the Appendix); for the moment you should of course assume that $\cos x$ is well defined and has its usual properties (e.g., it is 2π-periodic, has absolute value ≤ 1 and takes each value in $[-1, 1]$ exactly once if we restrict x to the interval $[0, \pi]$, etc.). Thus, $|\cos\theta| \le 1$ (used in the last step above) is certainly correct.

2.2 (a) The Cauchy-Schwarz inequality proved above guarantees $|\underline{x} \cdot \underline{y}| \le \|\underline{x}\| \, \|\underline{y}\|$ and hence $\underline{x} \cdot \underline{y} \le \|\underline{x}\| \, \|\underline{y}\|$ for all $\underline{x}, \underline{y} \in \mathbb{R}^n$. If $\underline{y} \ne \underline{0}$, prove that equality holds in the first inequality if and only if $\underline{x} = \lambda\underline{y}$ for some $\lambda \in \mathbb{R}$ and equality holds in the second inequality if and only if $\underline{x} = \lambda\underline{y}$ for some $\lambda \ge 0$.

(b) By examining the proof of the triangle inequality $\|\underline{x} + \underline{y}\| \le \|\underline{x}\| + \|\underline{y}\|$ given above (recall that proof began with the identity $\|\underline{x} + \underline{y}\|^2 = \|\underline{x}\|^2 + \|\underline{y}\|^2 + 2\underline{x} \cdot \underline{y}$), prove that *equality* holds in the triangle inequality $\Longleftrightarrow$ either at least one of $\underline{x}, \underline{y}$ is $\underline{0}$ or $\underline{x}, \underline{y} \ne \underline{0}$ and $\underline{y} = \lambda\underline{x}$ with $\lambda > 0$.

2.3 (Another proof of the Cauchy-Schwarz inequality.) If $\underline{a} = (a_1, \ldots, a_n)$, $\underline{b} = (b_1, \ldots, b_n) \in \mathbb{R}^n$, prove the identity $\frac{1}{2}\sum_{i,j=1}^{n}(a_i b_j - a_j b_i)^2 = \|\underline{a}\|^2 \|\underline{b}\|^2 - (\underline{a} \cdot \underline{b})^2$, and hence prove $|\underline{a} \cdot \underline{b}| \le \|\underline{a}\| \, \|\underline{b}\|$.

2.4 Using the dot product, prove, for any vectors $\underline{x}, \underline{y} \in \mathbb{R}^n$:

(a) The parallelogram law: $\|\underline{x} - \underline{y}\|^2 + \|\underline{x} + \underline{y}\|^2 = 2(\|\underline{x}\|^2 + \|\underline{y}\|^2)$.

(b) The law of cosines: $\|\underline{x} - \underline{y}\|^2 = \|\underline{x}\|^2 + \|\underline{y}\|^2 - 2\|\underline{x}\| \, \|\underline{y}\| \cos\theta$, assuming $\underline{x}, \underline{y}$ are nonzero and θ is the angle between $\underline{x}$ and $\underline{y}$.

(c) Give a geometric interpretation of identities (a),(b) (i.e., describe what (a) is saying about the parallelogram determined by $\underline{x}, \underline{y}$—i.e., $OACB$ where $\overrightarrow{OA} = \underline{x}$, $\overrightarrow{OB} = \underline{y}$, $\overrightarrow{OC} = \underline{x} + \underline{y}$, and what (b) is saying about the triangle determined by $\underline{x}$ and $\underline{y}$—i.e., OAB, where $\overrightarrow{OA} = \underline{x}$, $\overrightarrow{OB} = \underline{y}$).

3   SUBSPACES AND LINEAR DEPENDENCE OF VECTORS

We say that a subset V of $\mathbb{R}^n$ is a *subspace* of $\mathbb{R}^n$ (or more specifically a *linear subspace* of $\mathbb{R}^n$) if V has the properties that it contains at least the zero vector $\underline{0} = (0, \ldots, 0)$, i.e.,

3.1
$$\underline{0} \in V$$

and if it is *closed under addition and multiplication by scalars*, i.e.,

3.2
$$\underline{x}, \underline{y} \in V \text{ and } \lambda, \mu \in \mathbb{R} \Rightarrow \lambda\underline{x} + \mu\underline{y} \in V.$$

For example, the subset V consisting of just the zero vector (i.e., $V = \{\underline{0}\}$) is a subspace (usually referred to as the *trivial subspace*) and the subset V consisting of all vectors in $\mathbb{R}^n$ (i.e., $V = \mathbb{R}^n$) is a subspace.

To facilitate discussion of some less trivial examples we need to introduce some important terminology, as follows: We first define *the span* of a given collection $\underline{v}_1, \ldots, \underline{v}_N$ of vectors in $\mathbb{R}^n$.

3.3 Definition:

$$\text{span}\{\underline{v}_1, \ldots, \underline{v}_N\} = \{c_1\underline{v}_1 + c_2\underline{v}_2 + \cdots c_N\underline{v}_N : c_1, \ldots, c_N \in \mathbb{R}\}\,.$$

An expression of the form $c_1\underline{v}_1 + c_2\underline{v}_2 + \cdots + c_N\underline{v}_N$ (with $c_1, \ldots, c_N \in \mathbb{R}$) is *a linear combination of* $\underline{v}_1, \ldots, \underline{v}_N$, and the linear combination is said to be *nontrivial* if not all the $c_1, \ldots, c_N$ are zero. Using this terminology the Def. 3.3 says that $\text{span}\{\underline{v}_1, \ldots, \underline{v}_N\}$ is the set of all linear combinations of $\underline{v}_1, \ldots, \underline{v}_N$.

We claim the following.

3.4 Lemma. *For any given nonempty collection* $\underline{v}_1, \ldots, \underline{v}_N \in \mathbb{R}^n$, $\text{span}\{\underline{v}_1, \ldots, \underline{v}_N\}$ *is a subspace of* $\mathbb{R}^n$.

Proof: Let $V = \text{span}\{\underline{v}_1, \ldots, \underline{v}_N\}$. Notice that $c_1\underline{v}_1 + \cdots + c_N\underline{v}_N = \underline{0}$ if $c_j = 0 \,\forall\, j = 1, \ldots, N$, hence $\underline{0} \in V$. Also, if $w_1, w_2 \in V$ then $\underline{w}_1 = c_1\underline{v}_1 + \cdots + c_N\underline{v}_N$ and $\underline{w}_2 = d_1\underline{v}_1 + \cdots + d_N\underline{v}_N$ for suitable choices of $c_j, d_j \in \mathbb{R}$, and so $\lambda\underline{w}_1 + \mu\underline{w}_2 = (\lambda c_1 + \mu d_1)\underline{v}_1 + \cdots + (\lambda c_N + \mu d_N)\underline{v}_N \in V$.

One of the theorems we'll prove later is that in fact *any* subspace V of $\mathbb{R}^n$ can be expressed as the span of a suitable collection of vectors $\underline{v}_1, \ldots, \underline{v}_N \in \mathbb{R}^n$.

We need one more piece of terminology.

3.5 Definition: Vectors $\underline{v}_1, \ldots, v_N$ are *linearly dependent* (l.d.) if some nontrivial linear combination of $\underline{v}_1, \ldots, \underline{v}_N$ is the zero vector, i.e., $\underline{v}_1, \ldots, \underline{v}_N$ are linearly dependent if there are constants $c_1, \ldots, c_N$ not all zero such that $c_1\underline{v}_1 + \cdots + c_N\underline{v}_N = \underline{0}$.

Notice that $\underline{v}_1, \ldots, \underline{v}_N$ are l.d. if and only if one of the vectors $\underline{v}_1, \ldots, \underline{v}_N$ can be expressed as a linear combination of the others. That is:

3.6 Lemma. $\underline{v}_1, \ldots, \underline{v}_N$ *are l.d.* $\iff \exists\, j \in \{1, \ldots, N\}$ *and constants* $c_i,\ i \neq j$, *with* $v_j = \sum_{i \neq j, i=1}^{N} c_i\underline{v}_i$.

Proof of $\Rightarrow$: We are given $c_1, \ldots, c_N$ not all zero such that $\sum_i c_i\underline{v}_i = \underline{0}$. Pick j such that $c_j \neq 0$. Then $\underline{v}_j = \sum_{i \neq j, i=1}^{N} \frac{-c_i}{c_j}\underline{v}_i$.

Proof of $\Leftarrow$: Assume $j \in \{1, \ldots, N\}$ and $c_i,\ i \neq j$, are such that $\underline{v}_j = \sum_{i \neq j, i=1}^{N} c_i\underline{v}_i$. Then $\sum_{i=1}^{N} c_i\underline{v}_i = \underline{0}$, where we define $c_j = -1$.

SECTION 3 EXERCISES

3.1 Let $\underline{v}_1, \ldots, \underline{v}_k$ be any set of vectors in $\mathbb{R}^n$. Let $\widetilde{\underline{v}}_1, \ldots, \widetilde{\underline{v}}_k$ be obtained by applying any one of the 3 "elementary operations" to $\underline{v}_1, \ldots, \underline{v}_k$, where the 3 elementary operations are: (i) interchange of any pair of vectors (i.e., (i) merely changes the ordering of the vectors); (ii) multiplication of one of the vectors by a nonzero scalar; (iii) replacing the i-th vector $\underline{v}_i$ by the new vector $\widetilde{\underline{v}}_i = \underline{v}_i + \mu \underline{v}_j$, where $i \neq j$ and $\mu \in \mathbb{R}$.) Prove that $\text{span}\{\underline{v}_1, \ldots, \underline{v}_k\} = \text{span}\{\widetilde{\underline{v}}_1, \ldots, \widetilde{\underline{v}}_k\}$.

4   GAUSSIAN ELIMINATION AND THE LINEAR DEPENDENCE LEMMA

We begin with some remarks about linear systems of equations:

$$(*) \quad \begin{cases} a_{11}x_1+ & a_{12}x_2+ & \cdots+ & a_{1n}x_n & = b_1 \\ a_{21}x_1+ & a_{22}x_2+ & \cdots+ & a_{2n}x_n & = b_2 \\ & \vdots & \vdots & \vdots & \\ a_{m1}x_1+ & a_{m2}x_2+ & \cdots+ & a_{mn}x_n & = b_m, \end{cases}$$

where a_{ij}, b_i are given real constants and the *unknowns* $x_1, \ldots, x_n$ are to be determined.

Terminology:

(1) Any choice $x_1, \ldots, x_n$ which satisfies all m of the equations is called a solution of the system.

(2) The corresponding vector $(x_1, x_2, \cdots, x_n)^T$ is called a vector solution (or a solution vector) for the system. (Here we use the notation that $(x_1, x_2, \cdots, x_n)^T$ denotes the column vector with j-th entry x_j; for reasons which will become apparent later, we always write solution vectors as column vectors rather than row vectors.)

(3) The set of all possible solution vectors is called the solution set of the system.

There is a systematic procedure for solving (i.e., finding the solution set of) such linear systems, known as Gaussian elimination. We'll discuss Gaussian elimination and its consequences in detail later in this chapter, but for the moment we just need to describe the first step in the process.

The procedure is based on the (easily checked) observation that the set of solutions of the system $(*)$ is unchanged under any one of the following 3 operations:

(i)   interchanging any two of the equations;

(ii)   multiplying any one of the equations by a nonzero constant;

(iii) adding a multiple of one equation to another one of the equations, i.e., if we add a multiple of the i-th equation to the j-th equation, where $i \neq j$.

We now consider the system $(*)$, and the following alternatives.

Case 1: The coefficient a_{i1} of x_1 is zero for each $i = 1, \ldots, m$; thus in this case the system $(*)$ has the form

$$\begin{cases} 0x_1+ & a_{12}x_2+ & \cdots+ & a_{1n}x_n & = b_1 \\ 0x_1+ & a_{22}x_2+ & \cdots+ & a_{2n}x_n & = b_2 \\ & \vdots & \vdots & \vdots & \\ 0x_1+ & a_{m2}x_2+ & \cdots+ & a_{mn}x_n & = b_m \end{cases}$$

(i.e., the unknown x_1 does not appear at all in any of the equations).

Case 2: At least one of the coefficients $a_{i1} \neq 0$; then by operation (i) we may arrange that $a_{11} \neq 0$, and by operation (ii) we may actually arrange that $a_{11} = 1$, so that the system (∗) can be changed to a new system having the same set of solutions as the original system and having the form

$$
\begin{cases}
1\ x_1 + & \tilde{a}_{12}x_2 + & \cdots + & \tilde{a}_{1n}x_n & = \tilde{b}_1 \\
\tilde{a}_{21}x_1 + & \tilde{a}_{22}x_2 + & \cdots + & \tilde{a}_{2n}x_n & = \tilde{b}_2 \\
& \vdots & \vdots & \vdots & \\
\tilde{a}_{m1}x_1 + & \tilde{a}_{m2}x_2 + & \cdots + & \tilde{a}_{mn}x_n & = \tilde{b}_m.
\end{cases}
$$

Now the operation (iii) allows us to subtract $\tilde{a}_{21}$ times the first equation from the second equation, $\tilde{a}_{31}$ times the first equation from the third equation, and so on, thus giving a new system having the same set of solutions as the original system and having the form

$$
\begin{cases}
1x_1 + & \hat{a}_{12}x_2 + & \cdots + & \hat{a}_{1n}x_n & = \hat{b}_1 \\
0x_1 + & \hat{a}_{22}x_2 + & \cdots + & \hat{a}_{2n}x_n & = \hat{b}_2 \\
& \vdots & \vdots & \vdots & \\
0x_1 + & \hat{a}_{m2}x_2 + & \cdots + & \hat{a}_{mn}x_n & = \hat{b}_m
\end{cases}
$$

(i.e., the x_1 unknown only appears in the first equation), where each $\hat{b}_j$ is a linear combination of the original $b_1, \ldots, b_n$; in particular, $\hat{b}_j = 0$ for each $j = 1, \ldots, m$ if all the original $b_j = 0$, $j = 1, \ldots, m$.

This completes the first step of the Gaussian elimination process. Notice that this first step provides us with a new system of equations which is *equivalent* to (i.e., has precisely the same solution set as) the original system, and which either has the form

(∗)′
$$
\begin{cases}
0x_1 + & a_{12}x_2 + & \cdots + & a_{1n}x_n & = b_1 \\
0x_1 + & a_{22}x_2 + & \cdots + & a_{2n}x_n & = b_2 \\
& \vdots & \vdots & \vdots & \\
0x_1 + & a_{m2}x_2 + & \cdots + & a_{mn}x_n & = b_m
\end{cases}
$$

(i.e., the x_1 unknown does not appear at all in any of the equations), or

(∗)″
$$
\begin{cases}
1x_1 + & \hat{a}_{12}x_2 + & \cdots + & \hat{a}_{1n}x_n & = \hat{b}_1 \\
0x_1 + & \hat{a}_{22}x_2 + & \cdots + & \hat{a}_{2n}x_n & = \hat{b}_2 \\
& \vdots & \vdots & \vdots & \\
0x_1 + & \hat{a}_{m2}x_2 + & \cdots + & \hat{a}_{mn}x_n & = \hat{b}_m
\end{cases}
$$

(i.e., the x_1 unknown only appears in the first equation).

The system (∗) is said to be homogeneous if all the $b_j = 0$, $j = 1, \ldots, m$; as we noted above, the first step in the Gaussian elimination process gives a new system which is also homogeneous if the original system is homogeneous.

The following lemma refers to such a homogeneous system of linear equations, i.e., a system of m equations in n unknowns thus:

$$(**) \quad \begin{cases} a_{11}x_1 + & a_{12}x_2 + & \cdots + & a_{1n}x_n & = 0 \\ a_{21}x_1 + & a_{22}x_2 + & \cdots + & a_{2n}x_n & = 0 \\ & \vdots & \vdots & \vdots & \\ a_{m1}x_1 + & a_{m2}x_2 + & \cdots + & a_{mn}x_n & = 0. \end{cases}$$

Such a system is called "under-determined" if there are fewer equations than unknowns, i.e., if $m < n$.

4.1 "Under-determined systems lemma:" *An under-determined homogeneous system of linear equations (i.e., a system as in (**) above with $m < n$) always has a nontrivial solution.*

Proof: The proof is based on induction on m. When $m = 1$ the given system is just the single equation

$$a_{11}x_1 + a_{12}x_2 + \cdots + a_{1n}x_n = 0$$

with $n \geq 2$, and if $a_{11} = 0$ we see that $x_1 = 1$, $x_2 = \cdots = x_n = 0$ is a nontrivial solution. On the other hand, if $a_{11} \neq 0$ then we get a nontrivial solution by taking $x_2 = 1$, $x_1 = -(a_{12}/a_{11})$, and $x_3 = \cdots = x_n = 0$. Thus, the lemma is true in the case $m = 1$.

So take $m \geq 2$ and, as an inductive hypothesis, *assume* that the theorem is true with $m - 1$ in place of m, and that we are given the homogeneous system (**) with $m < n$. According to the above discussion of Gaussian elimination we know that (**) has the same set of solutions as a system of the form

$$(**) \quad \begin{cases} \widehat{a}_{11}x_1 + & \widehat{a}_{12}x_2 + & \cdots + & \widehat{a}_{1n}x_n & = 0 \\ 0\,x_1 + & \widehat{a}_{22}x_2 + & \cdots + & \widehat{a}_{2n}x_n & = 0 \\ & \vdots & \vdots & \vdots & \\ 0\,x_1 + & \widehat{a}_{m2}x_2 + & \cdots + & \widehat{a}_{mn}x_n & = 0, \end{cases}$$

where either $\widehat{a}_{11} = 1$ or 0, so it suffices to show that there is a nontrivial solution of (**). The last $m - 1$ equations here is an under-determined system of $m - 1$ equations in the $n - 1$ unknowns $x_2, \ldots, x_n$, so by the inductive hypothesis there is a nontrivial solution $x_2, \ldots, x_n$. If $\widehat{a}_{11} = 1$ we also solve the first equation by taking $x_1 = -(\widehat{a}_{12}x_2 + \cdots + \widehat{a}_{1n}x_n)$, so the proof is complete in the case $\widehat{a}_{11} = 1$.

On the other hand, if $\widehat{a}_{11} = 0$ then we get a nontrivial solution of the system (**) by simply taking $x_1 = 1$ and $x_2 = \cdots = x_n = 0$, so the proof is complete.

A very important corollary of the under-determined systems lemma is the following linear dependence lemma.

4.2 "Linear Dependence Lemma." *Suppose $\underline{v}_1, \underline{v}_2, \ldots, \underline{v}_k$ are any vectors in $\mathbb{R}^n$. Then any $k + 1$ vectors in $\mathrm{span}\{\underline{v}_1, \ldots, \underline{v}_k\}$ are l.d.*

4.3 Remark: An important special case of this is when $k = n$ and $\underline{v}_j = \underline{e}_j$, where $\underline{e}_j$ is the j-th standard basis vector as in 1.4. Since $\mathrm{span}\{\underline{e}_1, \ldots, \underline{e}_n\} = $ all of $\mathbb{R}^n$ (by 1.4), in this case the above Linear Dependence Lemma says simply that any $n + 1$ vectors in $\mathbb{R}^n$ are linearly dependent.

Proof of the Linear Dependence Lemma: Let $\underline{w}_1, \ldots, \underline{w}_{k+1} \in \mathrm{span}\{\underline{v}_1, \ldots, \underline{v}_k\}$. Then each $\underline{w}_j = $ a linear combination of $\underline{v}_1, \ldots, \underline{v}_k$, so that

$$(1) \qquad \underline{w}_j = a_{1j}\underline{v}_1 + a_{2j}\underline{v}_2 + \cdots + a_{kj}\underline{v}_k, \quad j = 1, \ldots, k+1,$$

for suitable scalars a_{ij}, $i = 1, \ldots, k$, $j = 1, \ldots, k+1$.

Now we want to prove that $\underline{w}_1, \ldots, \underline{w}_{k+1}$ are l.d.; in other words, we want to prove that the system of equations

$$(2) \qquad x_1\underline{w}_1 + \cdots + x_{k+1}\underline{w}_{k+1} = \underline{0}$$

has a nontrivial solution (i.e., we must show that we can find $x_1, \ldots, x_{k+1}$ which are not all zero and which satisfy the system (2)).

Substituting the given expressions (1) for the $\underline{w}_j$ into (2), we see that (2) actually says

$$x_1(a_{11}\underline{v}_1 + \cdots + a_{k1}\underline{v}_k) + x_2(a_{12}\underline{v}_1 + \cdots + a_{k2}\underline{v}_k) + \cdots$$
$$\cdots + x_{k+1}(a_{1\,k+1}\underline{v}_1 + \cdots + a_{k\,k+1}\underline{v}_k) = \underline{0};$$

that is,

$$(3) \qquad (x_1 a_{11} + \cdots + x_{k+1}a_{1\,k+1})\underline{v}_1 + (x_1 a_{21} + \cdots + x_{k+1}a_{2\,k+1})\underline{v}_2 + \cdots$$
$$\cdots + (x_1 a_{k\,1} + x_2 a_{k\,2} + \cdots + x_{k+1}a_{k\,k+1})\underline{v}_k = \underline{0},$$

so it suffices to get a nontrivial solution of the homogeneous linear system

$$
\begin{aligned}
a_{11}x_1 + \quad a_{12}x_2 + \quad \cdots + \quad a_{1\,k+1}x_{k+1} &= 0 \\
a_{21}x_1 + \quad a_{22}x_2 + \quad \cdots + \quad a_{2\,k+1}x_{k+1} &= 0 \\
\vdots \qquad\quad \vdots \qquad\qquad \vdots \qquad\qquad \\
a_{k1}x_1 + \quad a_{k2}x_2 + \quad \cdots + \quad a_{k\,k+1}x_{k+1} &= 0
\end{aligned}
$$

of k equations in the $k + 1$ unknowns $x_1, \ldots, x_{k+1}$, and we can do this by the under-determined systems Lem. 4.1, so the proof is complete.

SECTION 4 EXERCISES

4.1 Use the Linear Dependence Lemma to prove that if $k \in \{2, \ldots, n\}$, then k vectors $\underline{v}_1, \ldots, \underline{v}_k \in \mathbb{R}^n$ are l.d. $\iff \dim \mathrm{span}\{\underline{v}_1, \ldots, \underline{v}_k\} \leq k - 1$.

5   THE BASIS THEOREM

We now introduce the notion of *basis* for a nontrivial subspace of $\mathbb{R}^n$. For this we first need the appropriate definitions, as follows.

5.1 Definition: Vectors $\underline{v}_1, \ldots, \underline{v}_N \in \mathbb{R}^n$ are *linearly independent* (l.i.) if they are not linearly dependent.

Thus, $\underline{v}_1, \ldots, \underline{v}_N$ are l.i. if and only if $\sum_{j=1}^{N} c_j \underline{v}_j = \underline{0} \Rightarrow c_j = 0 \, \forall \, j = 1, \ldots, N$.

5.2 Definition: Let V be any nontrivial subspace of $\mathbb{R}^n$. A *basis* for V is a collection of vectors $\underline{w}_1, \ldots, \underline{w}_q$ such that

(i) $\underline{w}_1, \ldots, \underline{w}_q$ are l.i., and
(ii) $\mathrm{span}\{\underline{w}_1, \ldots, \underline{w}_q\} = V$.

Such a basis always exists:

5.3 Theorem ("The Basis Theorem.") *Suppose V is a nontrivial subspace of $\mathbb{R}^n$. Then*

(a) *V has a basis; and*

(b) *If $\underline{u}_1, \ldots, \underline{u}_k$ are l.i. vectors in V then there is a basis for V which contains $\underline{u}_1, \ldots, \underline{u}_k$; more precisely, there is a basis $\underline{v}_1, \ldots, \underline{v}_q$ for V with $q \geq k$ and $\underline{v}_j = \underline{u}_j$ for each $j = 1, \ldots, k$.*

Proof of (a): Define

$$q = \max\{\ell \in \{1, \ldots, n\} : \exists \text{ l.i. vectors } \underline{w}_1, \ldots, \underline{w}_\ell \in V\},$$

and choose l.i. vectors $\underline{v}_1, \ldots, \underline{v}_q \in V$. (Thus, roughly speaking, $\underline{v}_1, \ldots, \underline{v}_q$ are chosen to give "a maximum number of linearly independent vectors in V.") We claim that such $\underline{v}_1, \ldots, \underline{v}_q$ must span V. To see this let $\underline{v}$ be an arbitrary vector in V and consider the vectors $\underline{v}_1, \ldots, \underline{v}_q, \underline{v}$. If $q = n$ this is a set of $n + 1$ vectors in $\mathbb{R}^n = \mathrm{span}\{\underline{e}_1, \ldots, \underline{e}_n\}$ and so must be l.d. by the Linear Dependence Lem. 4.2. On the other hand, if $q < n$ then $q + 1 \leq n$ and so $\underline{v}_1, \ldots, \underline{v}_q, \underline{v}$ must again be l.d., otherwise $\underline{v}_1, \ldots, \underline{v}_q, \underline{v}$ is a set of $q + 1 \in \{1, \ldots, n\}$ l.i. vectors in V contradicting the definition of q. Thus, in either case ($q = n, q < n$), the vectors $\underline{v}_1, \ldots, \underline{v}_q, \underline{v}$ are l.d. Thus, $c_0 \underline{v} + c_1 \underline{v}_1 + \cdots + c_q \underline{v}_q = \underline{0}$ for some $c_0, \ldots, c_q$ not all zero. But of course then $c_0 \neq 0$ because otherwise this identity would tell us that $c_1 \underline{v}_1 + \cdots + c_q \underline{v}_q = \underline{0}$ with not all $c_1, \ldots, c_q$ zero, contradicting the linear independence of $\underline{v}_1, \ldots, \underline{v}_q$. Thus, $\underline{v} = -c_0^{-1}(c_1 \underline{v}_1 + \cdots + c_q \underline{v}_q)$, which completes the proof.

Proof of (b): The proof of (b) is almost the same as the proof of (a), except that we start by defining

$$q = \max\{\ell \in \{k, \ldots, n\} : \exists \text{ l.i. vectors } \underline{w}_1, \ldots, \underline{w}_\ell \in V \\ \text{with } \underline{w}_j = \underline{u}_j \text{ for each } j = 1, \ldots, k\},$$

so that we can select l.i. vectors $\underline{v}_1, \ldots, \underline{v}_q \in V$ with $\underline{v}_j = \underline{u}_j$ for $j = 1, \ldots, k$. The remainder of the proof is identical, word for word, with the proof of (a), and yields the conclusion that $\underline{v}_1, \ldots, \underline{v}_q$ span V, and hence $\underline{v}_1, \ldots, \underline{v}_q$ are a basis for V with $q \geq k$ and $\underline{v}_j = \underline{u}_j$ for each $j = 1, \ldots, k$.

5.4 Theorem. *Let V be a nontrivial subspace of $\mathbb{R}^n$. Then every basis for V has the same number of vectors.*

Proof: (Based on the "Linear Dependence Lemma" 4.2.) Otherwise, we would have two sets of basis vectors $\underline{v}_1, \ldots, \underline{v}_k$ and $\underline{w}_1, \ldots, \underline{w}_q$ with $q > k$. Then, in particular, $V = \text{span}\{\underline{v}_1, \ldots, \underline{v}_k\} = \text{span}\{\underline{w}_1, \ldots, \underline{w}_q\} \supset \{\underline{w}_1, \ldots, \underline{w}_{k+1}\}$; that is, $\underline{w}_1, \ldots, \underline{w}_{k+1}$ are $k+1$ linearly independent vectors contained in $\text{span}\{\underline{v}_1, \ldots, \underline{v}_k\}$ which contradicts the linear dependence lemma.

In view of the above result we can now define the notion of "dimension" of a subspace.

5.5 Definition: If V is a nontrivial subspace of $\mathbb{R}^n$ then the dimension of V is the number of vectors in a basis for V. The trivial subspace $\{\underline{0}\}$ is said to have dimension zero.

5.6 Remarks: Of course $\mathbb{R}^n$ itself has dimension n, because $\underline{e}_1, \ldots, \underline{e}_n$ (as in 1.4) are l.i. and span all of $\mathbb{R}^n$; i.e., $\underline{e}_1, \ldots, \underline{e}_n$ is a basis of $\mathbb{R}^n$ (which we call the "standard basis for $\mathbb{R}^n$"; the vector $\underline{e}_j$ is called the "j-th standard basis vector").

We now prove a third theorem, which is also, like Thm. 5.4 above, a consequence of the Linear Dependence Lemma.

5.7 Theorem. *Let V be a nontrivial subspace of $\mathbb{R}^n$ of dimension k (so $k \leq n$ because any $n+1$ vectors in $\mathbb{R}^n$ are l.d. by the Linear Dependence Lemma). Then:*

(a) *any k vectors $\underline{v}_1, \ldots, \underline{v}_k$ in V which span V must be l.i. (and hence form a basis for V); and*

(b) *any k l.i. vectors $\underline{v}_1, \ldots, \underline{v}_k$ in V must span V (and hence form a basis for V).*

Proof of (a): $\underline{v}_1, \ldots, \underline{v}_k$ l.d. means (by Lem. 3.6 of the present chapter) that some $\underline{v}_j$ is a linear combination of the remaining $\underline{v}_i$ and hence for some $j \in \{1, \ldots, k\}$ we have

$$\text{span}\{\underline{v}_1, \ldots, \underline{v}_k\}(= V) = \text{span}\{\underline{v}_1, \ldots, \underline{v}_{j-1}, \underline{v}_{j+1}, \ldots, \underline{v}_k\}$$

and so any basis of V consists of k l.i. vectors in $\text{span}\{v_1, \ldots, \underline{v}_{j-1}, \underline{v}_{j+1}, \ldots, \underline{v}_k\}$, contradicting the linear dependence lemma.

Proof of (b): Otherwise, there is a vector $\underline{v} \in V$ which is not in $\text{span}\{\underline{v}_1, \ldots, v_k\}$ and so $\underline{v}_1, \ldots, \underline{v}_k, \underline{v}$ are $(k+1)$ l.i. vectors (see Exercise 5.2 below) in the subspace $V = \text{span}\{w_1, \ldots, \underline{w}_k\}$, where $\underline{w}_1, \ldots, \underline{w}_k$ is any basis for V, contradicting the linear dependence lemma.

SECTION 5 EXERCISES

5.1 Use the basis theorem to prove that if V, W are subspaces of $\mathbb{R}^n$ with $V \subset W$ and $\dim V = \dim W$, then $V = W$.

5.2 If $\underline{v}_1, \ldots, \underline{v}_k$ are l.i. and $\underline{v} \notin \text{span}\{\underline{v}_1, \ldots, \underline{v}_k\}$, prove that $\underline{v}_1, \ldots, \underline{v}_k, \underline{v}$ are l.i.

6   MATRICES

We begin with some basic definitions.

6.1 Definitions: (i) An $m \times n$ real matrix $A = (a_{ij})$ is an array of mn real numbers a_{ij}, arranged in m rows and n columns, with a_{ij} denoting the entry in the i-th row and j-th column. Thus, the i-th row is the vector

$$\rho_i = (a_{i1}, \ldots, a_{in}) \text{ (so } \rho_i^T \in \mathbb{R}^n)$$

and the j-th column is the vector

$$\underline{\alpha}_j = (a_{1j}, \ldots, a_{mj})^T \in \mathbb{R}^m .$$

(ii) If $A = (a_{ij})$ an $m \times n$ matrix as in (i) with columns $\underline{\alpha}_1, \ldots, \underline{\alpha}_n \in \mathbb{R}^m$ and if $\underline{x} = (x_1, \ldots, x_n)^T \in \mathbb{R}^n$, we define the matrix product $A\underline{x}$ to be the vector $\underline{y} = (y_1, \ldots, y_m)^T \in \mathbb{R}^m$ with

$$\underline{y} = \sum_{k=1}^{n} x_k \underline{\alpha}_k .$$

Equivalently,

$$y_i = \sum_{k=1}^{n} a_{ik} x_k, \quad i = 1, \ldots, m ,$$

which is also equivalent to

$$\underline{y} = \sum_{i=1}^{m} \sum_{k=1}^{n} a_{ik} x_k \underline{e}_i ,$$

where $\underline{e}_i$ is the i-th standard basis vector in $\mathbb{R}^m$ as in 1.4. We note, in particular, that by taking the special choice $\underline{x} = \underline{e}_j$ we see that this says exactly

$$A\underline{e}_j = \text{the } j\text{-th column } \underline{\alpha}_j = (a_{1j}, \ldots, a_{mj})^T = \sum_{i=1}^{m} a_{ij} \underline{e}_i \text{ of } A, \quad j = 1, \ldots, n .$$

(iii) More generally, if $A = (a_{ij})$ is an $m \times n$ matrix and $B = (b_{ij})$ is and $n \times p$ matrix, then AB denotes the $m \times p$ matrix with j-th column $A\underline{\beta}_j (\in \mathbb{R}^m)$, where $\underline{\beta}_j (\in \mathbb{R}^n)$ is the j-th column of B; thus

$$j\text{-th column of } AB = A\underline{\beta}_j \text{ where } \underline{\beta}_j = \text{ the } j\text{-th column of } B .$$

Equivalently,

$$AB = (c_{ij}), \text{ where } c_{ij} = \sum_{k=1}^{n} a_{ik} b_{kj}, \quad i = 1, \ldots, m, \ j = 1, \ldots, p .$$

(iv) If $A = (a_{ij})$ is an $m \times n$ matrix, then the *transpose* of A, denoted A^T, is the $n \times m$ matrix with entry a_{ji} in the i-th row and j-th column. Notice that this notation is consistent with our previous usage of the notation $\underline{x}^T$ when $\underline{x}$ is a row or column vector: If we are writing vectors in $\mathbb{R}^n$ as columns, then $\underline{x}^T$ denotes the row vector (i.e., $1 \times n$ matrix) with the same entries as $\underline{x}$ and if we are writing vectors in $\mathbb{R}^n$ as rows then $\underline{x}^T$ denotes the column (i.e., $n \times 1$ matrix) with the same entries as $\underline{x}$.

Notice that

$$\underline{\alpha}_j = j\text{-th column of } A \Rightarrow \underline{\alpha}_j^T \text{ is the } j\text{-th row of } A^T$$
$$\underline{\rho}_i = i\text{-th row of } A \Rightarrow \underline{\rho}_i^T \text{ is the } i\text{-th column of } A^T .$$

Notice that in a more formal sense an $m \times n$ matrix A can be thought of as a point in $\mathbb{R}^{nm}$, with the agreement that the entries are ordered into rows and columns rather than a single row or single column. From this point of view it is natural to define the sum of matrices, multiplication of a matrix by a scalar, and the *length* (or "*norm*") $\|A\|$ of a matrix, as in the following.

Analogous to addition of vectors in $\mathbb{R}^n$ we also add matrices *component-wise*; thus, if $A = (a_{ij})$, $B = (b_{ij})$ are $m \times n$ matrices then we define $A + B$ to the be $m \times n$ matrix with $a_{ij} + b_{ij}$ in the i-th row and j-th column. Thus,

6.2 $$(a_{ij}) + (b_{ij}) = (a_{ij} + b_{ij}) .$$

Note, however, that it does not make sense to add matrices of different sizes.

Again analogous to the corresponding operation for vectors in $\mathbb{R}^n$, for a given $m \times n$ matrix $A = (a_{ij})$ and a given $\lambda \in \mathbb{R}$ we define

6.3 $$\lambda A = (\lambda a_{ij}) ,$$

and for such an $m \times n$ matrix $A = (a_{ij})$ we define the "length" or "norm" of A, $\|A\|$, by

6.4 $$\|A\| = \sqrt{\sum_{i=1}^{m} \sum_{j=1}^{n} a_{ij}^2} \left(= \sqrt{\sum_{j=1}^{n} \sum_{i=1}^{m} a_{ij}^2} \right) .$$

Since after all this norm really is just the length of the vector with the nm entries a_{ij}, we can then use the usual properties of the norm of vectors in Euclidean space:

6.5 $$\|A + B\| \le \|A\| + \|B\|, \quad \|\lambda A\| = |\lambda| \|A\| .$$

6.6 Remark: Observe that if $A = (a_{ij})$ is $m \times n$ and $\underline{x} \in \mathbb{R}^n$ then the matrix product of A times $\underline{x}$ gives a vector $A\underline{x}$ in $\mathbb{R}^m$, *provided we agree to write vectors in $\mathbb{R}^n$ and $\mathbb{R}^m$ as columns*. Indeed, if $\underline{x} = (x_1, \ldots, x_n)$ is a row vector of length n (i.e., a $1 \times n$ matrix) with $n \ge 2$ then $A\underline{x}$ makes no sense. So we shall henceforth always assume we are writing vectors in $\mathbb{R}^n$ as columns when we wish to use the matrix product $A\underline{x}$ of an $m \times n$ matrix A by a vector in $\mathbb{R}^n$.

We also have the important inequality

6.7 $\|A\underline{x}\| \le \|A\|\|\underline{x}\|$, A any $m \times n$ matrix and $\underline{x} \in \mathbb{R}^n$.

Proof of 6.7: Since $\underline{x} = \sum_{j=1}^n x_j \underline{e}_j$ we have $A\underline{x} = \sum_{j=1}^n x_j A\underline{e}_j = \sum_{j=1}^n x_j \underline{\alpha}_j$, where $\underline{\alpha}_j$ is the j-th column of A (by (ii) above), and so, by the triangle inequality 2.6, $\|A\underline{x}\| \le \sum_{j=1}^n |x_j|\|\underline{\alpha}_j\| = \underline{y} \cdot \underline{\gamma}$, where $\underline{y} = (|x_1|, \ldots, |x_n|)^{\mathrm{T}}$, $\underline{\gamma} = (\|\underline{\alpha}_1\|, \ldots, \|\underline{\alpha}_n\|)^{\mathrm{T}} \in \mathbb{R}^n$, and hence by the Cauchy-Schwarz inequality we have

$$\|A\underline{x}\| \le \|\underline{y}\| \sqrt{\sum_{j=1}^n \|\underline{\alpha}_j\|^2} = \|\underline{x}\|\|A\|$$

as claimed.

One of the key reasons that matrices are important is that $m \times n$ matrices naturally represent linear transformations $\mathbb{R}^n \to \mathbb{R}^m$ as follows.

Suppose $T : \mathbb{R}^n \to \mathbb{R}^m$ is a linear transformation (i.e., T is a mapping from $\mathbb{R}^n$ to $\mathbb{R}^m$ such that $\lambda, \mu \in \mathbb{R}$, $\underline{x}, \underline{y} \in \mathbb{R}^m \Rightarrow T(\lambda\underline{x} + \mu\underline{y}) = \lambda T(\underline{x}) + \mu T(\underline{y})$). Then we can write $\underline{x} = \sum_{j=1}^n x_j \underline{e}_j$ and use the linearity of T to give $T(\underline{x}) = \sum_{j=1}^n x_j T(\underline{e}_j)$. Thus, if we let A be the $m \times n$ matrix with j-th column $\underline{\alpha}_j = T(\underline{e}_j)(\in \mathbb{R}^m)$ we then have by 6.1(ii) above that $T(\underline{x}) = A\underline{x}$. That is:

6.8 $T : \mathbb{R}^n \to \mathbb{R}^m$ linear $\Rightarrow T(\underline{x}) \equiv A\underline{x}$, where the j-th column of $A = T(\underline{e}_j)$.

SECTION 6 EXERCISES

6.1 Let $\theta \in [0, 2\pi)$ and let T be the linear transformation of $\mathbb{R}^2$ defined by $T(\underline{x}) = Q(\theta)\underline{x}$, where $Q(\theta)$ is the 2×2 matrix $\begin{pmatrix} \cos\theta & -\sin\theta \\ \sin\theta & \cos\theta \end{pmatrix}$. Prove that if $\underline{x} = \begin{pmatrix} r\cos\alpha \\ r\sin\alpha \end{pmatrix}$ (with $r \ge 0$ and $\alpha \in [0, 2\pi)$) then $T(\underline{x}) = \begin{pmatrix} r\cos(\alpha + \theta) \\ r\sin(\alpha + \theta) \end{pmatrix}$. With the aid of a sketch, give a geometric interpretation of this.

6.2 What is the matrix (in the sense described above) of the linear transformation $T : \mathbb{R}^2 \to \mathbb{R}^2$ which takes the point (x, y) to its "reflection in the line $y = x$," i.e., the transformation $T(x, y) = (y, x)$.

Caution: In this exercise points in $\mathbb{R}^2$ are written as row vectors, but in order to represent T in terms of matrix multiplication you should first rewrite everything in terms of column vectors.

7   RANK AND THE RANK-NULLITY THEOREM

Let $A = (a_{ij})$ be an $m \times n$ matrix. We define the column space $C(A)$ of A to be the subspace of $\mathbb{R}^m$ spanned by the columns of A: thus, if $\underline{\alpha}_j = (a_{1j}, a_{2j}, \ldots, a_{mj})^{\mathrm{T}}$ is the jth column of A, then

7.1 $C(A) = \text{span}\{\underline{\alpha}_1, \ldots, \underline{\alpha}_n\}$, which (by 3.4 of the present chapter) is a subspace of $\mathbb{R}^m$. The "null space" $N(A)$ of A is defined to be the set of all solutions $\underline{x} \in \mathbb{R}^n$ of the homogeneous system $A\underline{x} = \underline{0}$:

7.2
$$N(A) = \{\underline{x} \in \mathbb{R}^n : A\underline{x} = \underline{0}\} ,$$

so that $N(A)$ is a subspace of $\mathbb{R}^n$.

The *rank* (or column rank) of A is defined to be the dimension of the column space $C(A)$ of A; that is, it is the dimension of the subspace of $\mathbb{R}^n$ spanned by the columns of A.

The *row rank* of A is defined to be dimension of the subspace of $\mathbb{R}^n$ spanned by the *rows* of A, in case we agree to write all vectors in $\mathbb{R}^n$ as row vectors. (Equivalently, it is the column rank, i.e., the dimension of the column space, of A^{T}.)

One of the things we'll prove later (in Thm. 8.5 below) is that *the row rank and the column rank are equal*. This is not intuitively obvious, but true nevertheless.

The *nullity* of A is defined to be the dimension of the null space $N(A)$.

Let's look at the extreme cases for $N(A)$: (a) $N(A) = \{\underline{0}\}$ and (b) $N(A) = \mathbb{R}^n$ (i.e., the case when $N(A)$ is trivial and the case when $N(A)$ is the whole space $\mathbb{R}^n$). In case (a), we have that $A\underline{x} = \underline{0}$ has only the trivial solution $\underline{x} = \underline{0}$; but $A\underline{x} = \sum_{j=1}^{n} x_j \underline{\alpha}_j$, so this just says $\sum_{j=1}^{n} x_j \underline{\alpha}_j = \underline{0}$ has only the trivial solution $\underline{x} = \underline{0}$. That is, (a) just says that the columns of A are l.i., and hence are a basis for $C(A)$ and the column space $C(A)$ has dimension n. In case (b), we have $A\underline{x} = \underline{0}$ for each $\underline{x}$ in $\mathbb{R}^n$, which, in particular, means that $A\underline{e}_j = \underline{0}$, which says that the j-th column of A is the zero vector for each j, i.e., $A = O$ (the zero matrix, with every entry $= 0$), and hence $C(A) = \{\underline{0}\}$ in this case. Notice that in each of these two extreme cases we have $\dim C(A) + \dim N(A) = n$.

The rank/nullity theorem (below) says that this is in fact true in all cases, not just in the extreme cases.

7.3 Rank/Nullity Theorem. *Let A be any $m \times n$ matrix. Then*

$$\dim C(A) + \dim N(A) = n .$$

7.4 Remarks: (1) Notice this is correct in the extreme cases $N(A) = \{\underline{0}\}$ and $N(A) = \mathbb{R}^n$ (as we discussed above), so we only of have to give the proof in the nonextreme cases (i.e., when $N(A) \neq \{\underline{0}\}$ and $N(A) \neq \mathbb{R}^n$).

(2) Recall that $\dim C(A)$ is called the "rank" of A (i.e., "rank(A)") and $\dim N(A)$ the "nullity" of A (i.e., "nullity(A)"); using this terminology the above theorem says

$$\mathrm{rank}\,(A) + \mathrm{nullity}\,(A) = n ,$$

hence the name "rank/nullity theorem."

Proof of Rank/Nullity Theorem: By Rem. (1) above we can assume that $N(A) \neq \{\underline{0}\}$ and $N(A) \neq \mathbb{R}^n$. In particular, since $N(A)$ is nontrivial it has a basis $\underline{u}_1, \ldots, \underline{u}_k$. By the part (b) of the Basis

Thm. 5.3 we can find a basis $\underline{v}_1, \ldots, \underline{v}_n$ for all of $\mathbb{R}^n$ with $\underline{v}_j = \underline{u}_j$ for each $j = 1, \ldots, k$, and of course also $k \leq n - 1$ because $N(A) \neq \mathbb{R}^n$. Now

$$
\begin{aligned}
C(A) &= \mathrm{span}\{\underline{\alpha}_1, \ldots, \underline{\alpha}_n\} \text{ where } \underline{\alpha}_1, \ldots, \underline{\alpha}_n \text{ denote the columns of } A \\
&= \{x_1\underline{\alpha}_1 + \cdots + x_n\underline{\alpha}_n : x_1, \ldots, x_n \in \mathbb{R}\} \\
&= \{A\underline{x} : \underline{x} \in \mathbb{R}^n\} \\
&= \{A(\textstyle\sum_{j=1}^n c_j\underline{v}_j) : c_1, \ldots, c_n \in \mathbb{R}\} \text{ (because } \mathrm{span}\{\underline{v}_1, \ldots, \underline{v}_n\} = \mathbb{R}^n) \\
&= \{\textstyle\sum_{j=1}^n c_j A\underline{v}_j : c_1, \ldots, c_n \in \mathbb{R}\} \\
&= \{\textstyle\sum_{j=k+1}^n c_j A\underline{v}_j : c_{k+1}, \ldots, c_n \in \mathbb{R}\} \text{ (because } A\underline{v}_j = \underline{0} \text{ for } j = 1, \ldots, k) \\
&= \mathrm{span}\{A\underline{v}_{k+1}, \ldots, A\underline{v}_n\}\,.
\end{aligned}
$$

Also, $A\underline{v}_{k+1}, \ldots, A\underline{v}_n$ are l.i.; check:

$$
\begin{aligned}
\textstyle\sum_{j=k+1}^n c_j A\underline{v}_j = \underline{0} &\Rightarrow A\big(\textstyle\sum_{j=k+1}^n c_j\underline{v}_j\big) = \underline{0} \\
&\Rightarrow \textstyle\sum_{j=k+1}^n c_j\underline{v}_j \in N(A) = \mathrm{span}\{\underline{v}_1, \ldots, \underline{v}_k\} \\
&\Rightarrow \textstyle\sum_{j=k+1}^n c_j\underline{v}_j = \textstyle\sum_{j=1}^k d_j\underline{v}_j \text{ for some } d_1, \ldots, d_k \\
&\Rightarrow \textstyle\sum_{j=1}^k d_j\underline{v}_j - \textstyle\sum_{j=k+1}^n c_j\underline{v}_j = \underline{0} \\
&\Rightarrow c_{k+1}, \ldots, c_n, d_1, \ldots, d_k \text{ are all 0, because } \underline{v}_1, \ldots, \underline{v}_n \text{ are l.i.}
\end{aligned}
$$

Thus, $A\underline{v}_{k+1}, \ldots, A\underline{v}_n$ are a basis for $C(A)$, and hence $\dim C(A) = n - k = n - \dim N(A)$.

SECTION 7 EXERCISES

7.1 (a) Use Gaussian elimination to show that the solution set of the homogeneous system $A\underline{x} = \underline{0}$ with matrix

$$
A = \begin{pmatrix} 1 & 2 & 3 & 4 \\ 1 & 4 & 3 & 2 \\ 2 & 5 & 6 & 7 \\ 1 & 0 & 3 & 6 \end{pmatrix}
$$

is a plane (i.e., the span of 2 l.i. vectors), and find explicitly 2 l.i. vectors whose span is the solution space.

(b) Find the dimension of the column space of the above matrix (i.e., the dimension of the subspace of $\mathbb{R}^4$ which is spanned by the columns of the matrix).

Hint: Use the result of (a) and the rank nullity theorem.

7.2 If A, B are any $m \times n$ matrices, prove that $\mathrm{rank}(A + B) \leq \mathrm{rank}\,(A) + \mathrm{rank}\,(B)$.

($\mathrm{rank}\,(A)$ is the dimension of the column space of A, i.e., the dimension of the subspace of $\mathbb{R}^m$ spanned by the columns of A.)

7.3 (a) If A, B are, respectively, $m \times n$ and $n \times p$ matrices, prove that $\mathrm{rank}\,AB \leq \min\{\mathrm{rank}\,A, \mathrm{rank}\,B\}$.

(b) Give an example in the case $n = m = p = 2$ to show that strict inequality may hold in the inequality of (a).

7.4 Suppose $\underline{v}_1, \ldots, \underline{v}_n$ are vectors in $\mathbb{R}^m$, not all zero, and let $\ell \in \{1, \ldots, n\}$ be the maximum number of linearly independent vectors which can be selected from $\underline{v}_1, \ldots, \underline{v}_n$. Select any $1 \le j_1 < j_2 < \cdots < j_\ell \le n$ such that $\underline{v}_{j_1}, \ldots, \underline{v}_{j_\ell}$ are l.i. Prove that $\underline{v}_{j_1}, \ldots, \underline{v}_{j_\ell}$ is a basis for $\mathrm{span}\{\underline{v}_1, \ldots, \underline{v}_n\}$.

(In particular, given a nonzero $m \times n$ matrix A with j-th column $\underline{\alpha}_j$, we can always select certain of the columns $\underline{\alpha}_{j_1}, \ldots, \underline{\alpha}_{j_Q}$ to give a basis for $C(A)$.)

8   ORTHOGONAL COMPLEMENTS AND ORTHOGONAL PROJECTION

Let V be a subspace of $\mathbb{R}^n$. The "orthogonal complement" $V^\perp$ of V is defined to be the set of all vectors $\underline{x}$ in $\mathbb{R}^n$ such that $\underline{x} \cdot \underline{v} = 0$ for each $\underline{v} \in V$. That is, $V^\perp$ is the set of all vectors which are orthogonal to each vector in V. Of course $V^\perp$ so defined is a subspace of $\mathbb{R}^n$.

If V is the trivial subspace $\{\underline{0}\}$ we evidently have $V^\perp = \mathbb{R}^n$, whereas if V is nontrivial the Basis Thm. 5.3(a) guarantees that we can choose a basis $\underline{u}_1, \ldots, \underline{u}_k$ for V and then by 5.3(b) (applied with $\mathbb{R}^n$ in place of V) there is a basis $\underline{v}_1, \ldots, \underline{v}_n$ for $\mathbb{R}^n$ with $\underline{v}_j = \underline{u}_j$ for each $j = 1, \ldots, k$. It is then very easy to check that $V^\perp$ is exactly characterized by saying that

$$\underline{x} \in V^\perp \iff \underline{u}_j \cdot \underline{x} = 0 \text{ for each } j = 1, \ldots, k .$$

The k equations here form a homogeneous system of k linear equations in the n unknowns $x_1, \ldots, x_n$, so by the under-determined systems Lem. 4.1 it is a *nontrivial* subspace unless $k = n$, i.e., unless $V = \mathbb{R}^n$.

Our main initial results about $V^\perp$ are given in the following theorem.

8.1 Theorem. *Let V be a subspace of $\mathbb{R}^n$. Then the subspace $V^\perp$ satisfies the following:*

(i) $$V^\perp \cap V = \{\underline{0}\}$$
(ii) $$V + V^\perp = \mathbb{R}^n$$
(iii) $$\dim V + \dim V^\perp = n$$
(iv) $$(V^\perp)^\perp = V .$$

Remark: Notice that (i) says that V and $V^\perp$ have only the zero vector in common, and (ii) says that every vector in $\mathbb{R}^n$ can be written as a the sum of a vector in V and a vector in $V^\perp$.

Proof of (i): $\underline{w} \in V \cap V^\perp \Rightarrow \underline{w} \in V$ and $\underline{w} \in V^\perp$. Thus, in particular, $\underline{w} \cdot \underline{w} = 0$, i.e., $\|\underline{w}\|^2 = 0$, i.e., $\underline{w} = \underline{0}$.

Proof of (ii): $V + V^\perp = \{\underline{v} + \underline{u} : \underline{v} \in V, \underline{u} \in V^\perp\}$ and this is clearly a subspace of $\mathbb{R}^n$. Call this subspace W. Take any vector $\underline{x} \in W^\perp$, so that $\underline{x} \cdot (\underline{v} + \underline{u}) = 0$ for each $\underline{v} \in V$ and each $\underline{u} \in V^\perp$.

Taking $\underline{u} = \underline{0}$, this implies $\underline{x} \cdot \underline{v} = 0$ for each $\underline{v} \in V$, so $\underline{x} \in V^{\perp}$, and, taking $\underline{v} = \underline{0}$, it implies $\underline{x} \cdot \underline{u} = 0$ for each $\underline{u} \in V^{\perp}$. In particular, since $\underline{x} \in V^{\perp}$, we must have $\underline{x} \cdot \underline{x} = 0$. Thus, $\underline{x} = \underline{0}$; that is, we have shown that $W^{\perp}$ is the trivial subspace, and, by the remarks preceding the statement of the theorem, this can occur only if $W = \mathbb{R}^n$, so (ii) is proved.

Proof of (iii): Notice that (iii) is trivially true if V is the trivial subspace (because then $V^{\perp} = \mathbb{R}^n$), so we can assume V is nontrivial. Likewise, we can assume that $V^{\perp}$ is nontrivial, because, as we pointed out in the discussion preceding the statement of the theorem, $V^{\perp}$ trivial implies $V = \mathbb{R}^n$ and again (iii) holds. Thus, we can assume that both V and $V^{\perp}$ are nontrivial, and hence by the basis theorem we can select a basis $\underline{v}_1, \ldots, \underline{v}_k$ for V and a basis $\underline{u}_1, \ldots, \underline{u}_\ell$ for $V^{\perp}$. Of course then by (ii) we have $\mathrm{span}\{\underline{v}_1, \ldots, \underline{v}_k, \underline{u}_1, \ldots, \underline{u}_\ell\} = \mathbb{R}^n$ and we claim that $\underline{v}_1, \ldots, \underline{v}_k, \underline{u}_1, \ldots, \underline{u}_\ell$ are linearly independent. To check this, suppose that $c_1 \underline{v}_1 + \cdots + c_k \underline{v}_k + d_1 \underline{u}_1 + \cdots + d_\ell \underline{u}_\ell = \underline{0}$. Then $c_1 \underline{v}_1 + \cdots + c_k \underline{v}_k = -d_1 \underline{u}_1 - \cdots - d_\ell \underline{u}_\ell$ and the left side is in V while the right side is in $V^{\perp}$, so by (i) both sides must be zero. That is, $c_1 \underline{v}_1 + \cdots + c_k \underline{v}_k = \underline{0} = d_1 \underline{u}_1 + \cdots + d_\ell \underline{u}_\ell$ and since $\underline{v}_1, \ldots, \underline{v}_k$ are l.i., this gives all the $c_j = 0$, and similarly, since $\underline{u}_1, \ldots, \underline{u}_\ell$ are l.i., all the $d_j = 0$. Thus, we have proved that the vectors $\underline{v}_1, \ldots, \underline{v}_k, \underline{u}_1, \ldots, \underline{u}_\ell$ both span $\mathbb{R}^n$ and are l.i.; that is, they are a basis for $\mathbb{R}^n$, and hence $k + \ell = n$. That is, $\dim V + \dim V^{\perp} = n$, as claimed.

Proof of (iv): Observe that by definition of $V^{\perp}$ we have $\underline{v} \cdot \underline{u} = 0$ for each $\underline{v} \in V$ and each $\underline{u} \in V^{\perp}$, which, in particular, says that each $\underline{v} \in V$ is in $(V^{\perp})^{\perp}$ (because it says that each $\underline{v} \in V$ is orthogonal to each vector in $V^{\perp}$), so we have shown $V \subset (V^{\perp})^{\perp}$.

By (iii) $\dim V^{\perp} = n - \dim V$. Also, $V^{\perp}$ is a subspace of $\mathbb{R}^n$, so (iii) holds with $V^{\perp}$ in place of V. That is, $\dim(V^{\perp})^{\perp} = n - \dim V^{\perp} = n - (n - \dim V) = \dim V$.

That is, we have shown that $V \subset (V^{\perp})^{\perp}$ and that $V, (V^{\perp})^{\perp}$ have the same dimension. They therefore must be equal. (Notice that here we use the following general principle: if V, W are subspaces of $\mathbb{R}^n$ and $V \subset W$ then $\dim V = \dim W$ implies $V = W$. This is an easy consequence of the Basis Thm. 5.3—see Exercise 5.1 above.)

This completes the proof of Thm. 8.1.

8.2 Remark: 8.1(ii) tells us that we can write any vector $\underline{x} \in \mathbb{R}^n$ as $\underline{x} = \underline{v} + \underline{u}$ with $\underline{v} \in V$ and $\underline{u} \in V^{\perp}$. We claim that such $\underline{v}, \underline{u}$ are *unique*. Indeed, if we could also write $\underline{x} = \widetilde{\underline{v}} + \widetilde{\underline{u}}$ with $\widetilde{\underline{v}} \in V$ and $\widetilde{\underline{u}} \in V^{\perp}$ then we would have $\underline{x} = \underline{v} + \underline{u} = \widetilde{\underline{v}} + \widetilde{\underline{u}}$, hence $\underline{v} - \widetilde{\underline{v}} = \widetilde{\underline{u}} - \underline{u}$. Since the left side is in V and the right side is in $V^{\perp}$, we would then have by 8.1(i) that $\underline{v} - \widetilde{\underline{v}} = \underline{0} = \widetilde{\underline{u}} - \underline{u}$, i.e., that $\underline{v} = \widetilde{\underline{v}}$ and $\underline{u} = \widetilde{\underline{u}}$. So $\underline{v}, \underline{u}$ are unique as claimed. For this reason the decomposition of 8.1(ii) is sometimes referred to as a *direct sum*.

Using Thm. 8.1 we can prove that existence of a unique *orthogonal projection* of $\mathbb{R}^n$ onto a given subspace V, as follows.

8.3 Theorem (Orthogonal Projection.) *Given a subspace $V \subset \mathbb{R}^n$ there is a unique map $P_V : \mathbb{R}^n \to \mathbb{R}^n$ with the two properties*

$$P_V(\underline{x}) \in V \quad \forall \underline{x} \in \mathbb{R}^n$$
$$\underline{x} - P_V(\underline{x}) \in V^\perp \quad \forall \underline{x} \in \mathbb{R}^n .$$

This map automatically has the additional properties

(i) *$P_V : \mathbb{R}^n \to \mathbb{R}^n$ is linear (i.e., $P_V(\lambda\underline{x} + \mu\underline{y}) = \lambda P_V(\underline{x}) + \mu P_V(\underline{y}) \, \forall \underline{x}, \underline{y} \in \mathbb{R}^n$)*
(ii) *$\underline{x} \cdot P_V(\underline{y}) = P_V(\underline{x}) \cdot \underline{y} \, \forall \underline{x}, \underline{y} \in \mathbb{R}^n$*
(iii) *$\forall \underline{x} \in \mathbb{R}^n$, $\|\underline{x} - P_V(x)\| \le \|\underline{x} - \underline{y}\| \, \forall \underline{y} \in V$, with equality $\iff \underline{y} = P_V(\underline{x})$.*

8.4 Remark: The map P_V is called *the orthogonal projection of $\mathbb{R}^n$ onto V*. Notice that the properties (i), (ii) tell us that P_V is given by matrix multiplication by a *symmetric* matrix (p_{ij}); that is, $P_V(\underline{x}) = \sum_{i,j=1}^n p_{ij} x_j \underline{e}_i$, where $p_{ij} = p_{ji}$ for each $i, j = 1, \ldots, n$, and property (iii) says that $P_V(\underline{x})$ is the unique nearest point of V to $\underline{x}$, as in the following schematic diagram depicting the case $n = 3$, $\dim V = 2$.

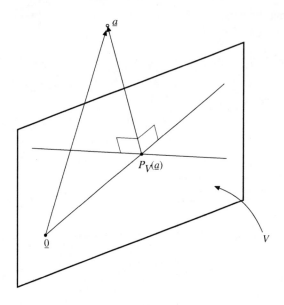

Figure 1.1: The orthogonal projection P_V onto a subspace V.

Proof of Theorem 8.3: Theorem 8.1(ii) implies that for any $\underline{x} \in \mathbb{R}^n$ we can write $\underline{x} = \underline{v} + \underline{u}$ with $\underline{v} \in V$ and $\underline{u} \in V^\perp$ and by Rem. 8.2 such $\underline{v}, \underline{u}$ are unique. Thus, we can *define* $P_V(\underline{x}) = \underline{v} (\in V)$, and then $\underline{x} - P_V(\underline{x}) = \underline{u} (\in V^\perp)$. Using the Thm. 8.1(i) we can readily check that the claimed uniqueness of P_V the stated properties (i) and (ii) (see Exercise 8.2 below).

To prove (iii), observe that for any $\underline{x} \in \mathbb{R}^n$ and $\underline{y} \in V$ we have $\|\underline{x} - \underline{y}\|^2 = \|(\underline{x} - P_V(\underline{x})) - (\underline{y} - P_V(\underline{x}))\|^2 = \|\underline{x} - P_V(\underline{x})\|^2 + \|\underline{y} - P_V(\underline{x})\|^2 + 2(\underline{x} - P_V(\underline{x})) \cdot (\underline{y} - P(\underline{x}))$, and the last term is zero because $x - P_V(x) \in V^\perp$ and $y - P_V(x) \in V$. That is, we have shown that $\|x - y\|^2 = \|x - P_V(x)\|^2 + \|y - P_V(x)\|^2 \geq \|x - P_V(x)\|^2$, with equality if and only if $y - P_V(x) = 0$, i.e., if and only if $y = P_V(x)$.

The following theorem, which is based on Thm. 8.1 above and the Rank-Nullity Theorem, establishes some important connections between column spaces and null spaces of a matrix A and its transpose A^T.

8.5 Theorem. *Let A be an $m \times n$ matrix. Then*

(i) $$(C(A))^\perp = N(A^T)$$
(ii) $$C(A) = (N(A^T))^\perp$$
(iii) $$\dim C(A) = \dim C(A^T) .$$

8.6 Remark: Notice that (iii) says that the row rank and the column rank are the same.

Proof of (i): Let $\underline{\alpha}_1, \ldots, \underline{\alpha}_n$ be the columns of A. Then

$$x \in (C(A))^\perp \iff x \cdot \underline{\alpha}_j = 0 \; \forall \, j = 1, \ldots, n$$
$$\iff A^T x = 0$$
$$\iff x \in N(A^T) .$$

Proof of (ii): By (i) $(C(A))^\perp = N(A^T)$ and hence (using (iv) of Thm. 8.1) $C(A) = ((C(A))^\perp)^\perp = (N(A^T))^\perp$.

Proof of (iii): By the rank-nullity theorem (applied to the transpose matrix A^T) we have

(1) $$\dim C(A^T) = m - \dim N(A^T) .$$

By part (iii) of Thm. 8.1 and by (i) above we have

(2) $$\dim C(A) = m - \dim(C(A))^\perp = m - \dim N(A^T) .$$

By (1),(2) we have $\dim C(A) = \dim C(A^T)$ as claimed.

SECTION 8 EXERCISES

8.1 (a) If V is a subspace of $\mathbb{R}^n$ and if P is the orthogonal projection onto V, prove that $I - P$ is the orthogonal projection onto $V^\perp$ (I is the identity transformation, so $(I - P)(x) = x - P(x)$).

(b) If $\underline{\alpha} \in \mathbb{R}^n$ with $\|\underline{\alpha}\| = 1$ and if $V = \text{span}\{\underline{\alpha}\}$, find the matrix of the orthogonal projection onto V and also the matrix of the orthogonal projection onto $V^\perp$.

8.2 If V and $P_V(x)$ are as in the first part of Thm. 8.3 above, prove that properties (i), (ii) hold as claimed.

Hint: To check (i) note that $P_V(\lambda x + \mu y) - \lambda P_V(x) - \mu P_V(y) = -(\lambda x + \mu y - P_V(\lambda x + \mu y)) + \lambda(x - P_V(x)) + \mu(y - P_V(y))$ and use Thm. 8.1(i) and the stated properties of P_V.

9   ROW ECHELON FORM OF A MATRIX

Let $A = (a_{ij})$ be an $m \times n$ matrix. Recall that the "First Stage" (described in Sec. 4 above) of the process of Gaussian elimination leads to 2 possible cases: either the first column is zero, or else there is at least one nonzero entry in the first column, and using only the 3 elementary row operations we produce a new matrix with all entries in the first column zero except for the first entry which we can arrange to be 1. That is, Stage 1 of the process of Gaussian elimination produces a new matrix $\widetilde{A}$ with the same null space as A (because elementary row operations leave the solution set of the homogeneous system $A\underline{x} = \underline{0}$ unchanged) and $\widetilde{A}$ either has the form:

Case 1
$$\widetilde{A} = \begin{pmatrix} 0 & \widetilde{a}_{12} & \cdots & \widetilde{a}_{1n} \\ 0 & \widetilde{a}_{22} & \cdots & \widetilde{a}_{2n} \\ \vdots & \vdots & \cdots & \vdots \\ 0 & \widetilde{a}_{m2} & \cdots & \widetilde{a}_{mn} \end{pmatrix}$$

or

Case 2
$$\widetilde{A} = \begin{pmatrix} 1 & \widetilde{a}_{12} & \cdots & \widetilde{a}_{1n} \\ 0 & \widetilde{a}_{22} & \cdots & \widetilde{a}_{2n} \\ \vdots & \vdots & \cdots & \vdots \\ 0 & \widetilde{a}_{m2} & \cdots & \widetilde{a}_{mn}. \end{pmatrix}$$

Now in Case 1 we can apply this whole first stage process again to the $m \times (n-1)$ matrix

$$\widehat{A} = \begin{pmatrix} \widetilde{a}_{12} & \cdots & \widetilde{a}_{1n} \\ \widetilde{a}_{22} & \cdots & \widetilde{a}_{2n} \\ \vdots & \cdots & \vdots \\ \widetilde{a}_{m2} & \cdots & \widetilde{a}_{mn}, \end{pmatrix}$$

while in Case 2 we can apply the whole first stage process again to the $(m-1) \times (n-1)$ matrix

$$\widehat{A} = \begin{pmatrix} \widetilde{a}_{22} & \cdots & \widetilde{a}_{2n} \\ \vdots & \cdots & \vdots \\ \widetilde{a}_{m2} & \cdots & \widetilde{a}_{mn} \end{pmatrix}.$$

Using mathematical induction on the number n of columns we can check that, in fact after at most n repetitions of the process described above, we obtain a matrix of the following form:

$$
\begin{array}{c}
\quad\quad\text{col. } j_1 \quad\quad\quad \text{col. } j_2 \quad\quad\quad \text{col. } j_3 \quad\cdots\quad \text{col. } j_Q \\
\left(\begin{array}{ccccccccccccccccc}
0 & \cdots & 0 & 1 & * & \cdots & * & * & * & \cdots & * & * & * & \cdots & * & * & * & \cdots & * \\
0 & \cdots & 0 & 0 & 0 & \cdots & 0 & 1 & * & \cdots & * & * & * & \cdots & * & * & * & \cdots & * \\
0 & \cdots & 0 & 0 & 0 & \cdots & 0 & 0 & 0 & \cdots & 0 & 1 & * & \cdots & * & * & * & \cdots & * \\
\vdots & & \vdots & \vdots & \vdots & & \vdots & \vdots & \vdots & & \vdots & \vdots & \vdots & & \vdots \\
0 & \cdots & 0 & 0 & 0 & \cdots & 0 & 0 & 0 & \cdots & 0 & 0 & 0 & \cdots & 0 & 1 & * & \cdots & * \\
0 & \cdots & 0 & 0 & 0 & \cdots & 0 & 0 & 0 & \cdots & 0 & 0 & 0 & \cdots & 0 & 0 & 0 & \cdots & 0 \\
\vdots & & \vdots & \vdots & \vdots & & \vdots & \vdots & \vdots & & \vdots & \vdots & \vdots & & \vdots \\
0 & \cdots & 0 & 0 & 0 & \cdots & 0 & 0 & 0 & \cdots & 0 & 0 & 0 & \cdots & 0
\end{array}\right)
\end{array}
$$

(a) — first group of rows; (b) — second group of rows

Schematic Diagram of the Row Echelon Form of A

where the first Q rows (labeled (a) above) are nonzero, and each contains a *pivot*, i.e., a 1 preceded by an unbroken string of zeros, and each successive one of these rows has a strictly longer unbroken string of zeros before the pivot; and the remaining rows (labeled (b) above), if there are any, are all zero.

Notice that mathematical induction does indeed provide a rigorous proof of this because the result is trivially true for $n = 1$, and for $n \geq 2$ the inductive hypothesis that the result is true for matrices with $n - 1$ columns would tell us that the matrix $\widehat{A}$ above can be reduced to echelon form by elementary row operations. By applying the same row operations to the matrix $\widetilde{A}$ we then get the required echelon form for A.

Such a form (obtained from A by elementary row operations) is called a *"row echelon form"* for the matrix A: Notice that there are a certain number $Q \leq \min\{m, n\}$ of nonzero rows, each of which consists of an unbroken string of zeros ("the leading zeros" of the row) followed by a 1 called the "leading 1" or "pivot," and each successive nonzero row has strictly more leading zeros than the previous. Correspondingly, the column numbers of the *pivot columns* (i.e., the columns which contain the pivots) are $j_1, \ldots, j_Q$, with $1 \leq j_1 < j_2 < \cdots < j_Q \leq n$, as shown in the above schematic diagram.

Of course by using further elementary row operations (subtracting the appropriate multiples of a row containing a pivot from the rows above it) we can now eliminate all nonzero entries *above* the pivot in each of the pivot columns, thus yielding a matrix called the "reduced row echelon form of A" (abbreviated rref A):

$$(\ddagger) \quad \begin{pmatrix} 0 & \cdots & 0 & 1 & * & \cdots & * & 0 & * & \cdots & * & 0 & * & \cdots & * & 0 & * & \cdots & * \\ 0 & \cdots & 0 & 0 & 0 & \cdots & 0 & 1 & * & \cdots & * & 0 & * & \cdots & * & 0 & * & \cdots & * \\ 0 & \cdots & 0 & 0 & 0 & \cdots & 0 & 0 & 0 & \cdots & 0 & 1 & * & \cdots & * & 0 & * & \cdots & * \\ \vdots & & \vdots & \vdots & \vdots & & \vdots & \vdots & \vdots & & \vdots & \vdots & \vdots & & \vdots & \vdots & \vdots & & \vdots \\ 0 & \cdots & 0 & 0 & 0 & \cdots & 0 & 0 & 0 & \cdots & 0 & 0 & 0 & \cdots & 0 & 1 & * & \cdots & * \\ 0 & \cdots & 0 & 0 & 0 & \cdots & 0 & 0 & 0 & \cdots & 0 & 0 & 0 & \cdots & 0 & 0 & 0 & \cdots & 0 \\ \vdots & & \vdots & \vdots & \vdots & & \vdots & \vdots & \vdots & & \vdots & \vdots & \vdots & & \vdots & \vdots & \vdots & & \vdots \\ 0 & \cdots & 0 & 0 & 0 & \cdots & 0 & 0 & 0 & \cdots & 0 & 0 & 0 & \cdots & 0 & 0 & 0 & \cdots & 0 \end{pmatrix}.$$

with columns indicated by col. j_1, col. j_2, col. j_3, $\cdots$, col. j_Q

The Reduced Row Echelon Form ("rref A") of A

Note: Such schematic diagrams are accurate and useful, but must be interpreted correctly; for instance, if $j_1 = 1$ (i.e., if there is a pivot in the first column), then the first few zero columns shown in the above diagrams are not present, and the reduced row echelon form of A (rref A) would look as follows:

$$\begin{pmatrix} 1 & * & \cdots & * & 0 & * & \cdots & * & 0 & * & \cdots & * & 0 & * & \cdots & * \\ 0 & 0 & \cdots & 0 & 1 & * & \cdots & * & 0 & * & \cdots & * & 0 & * & \cdots & * \\ 0 & 0 & \cdots & 0 & 0 & 0 & \cdots & 0 & 1 & * & \cdots & * & 0 & * & \cdots & * \\ \vdots & \vdots & & \vdots & \vdots & \vdots & & \vdots & \vdots & \vdots & & \vdots & \vdots & \vdots & & \vdots \\ 0 & 0 & \cdots & 0 & 0 & 0 & \cdots & 0 & 0 & 0 & \cdots & 0 & 1 & * & \cdots & * \\ 0 & 0 & \cdots & 0 & 0 & 0 & \cdots & 0 & 0 & 0 & \cdots & 0 & 0 & 0 & \cdots & 0 \\ \vdots & \vdots & & \vdots & \vdots & \vdots & & \vdots & \vdots & \vdots & & \vdots & \vdots & \vdots & & \vdots \\ 0 & 0 & \cdots & 0 & 0 & 0 & \cdots & 0 & 0 & 0 & \cdots & 0 & 0 & 0 & \cdots & 0 \end{pmatrix}.$$

with columns indicated by col. j_1, col. j_2, col. j_3, $\cdots$, col. j_Q

In the case when there is a pivot in every column (i.e., $Q = n$ and $j_1 = 1$, $j_2 = 2, \ldots, j_n = n$), a row echelon form of A would look like

$$\begin{pmatrix} 1 & * & * & \cdots & * \\ 0 & 1 & * & \cdots & * \\ \vdots & & \vdots & & \vdots \\ 0 & 0 & 0 & \cdots & 1 \\ 0 & 0 & 0 & \cdots & 0 \\ \vdots & & \vdots & & \vdots \\ 0 & 0 & 0 & \cdots & 0 \end{pmatrix}$$

and the reduced row echelon form of A (i.e., $\mathrm{rref}\,A$) is:

$(\ddagger\ddagger)$
$$\begin{pmatrix} 1 & 0 & 0 & \cdots & 0 \\ 0 & 1 & 0 & \cdots & 0 \\ \vdots & & \vdots & & \vdots \\ 0 & 0 & 0 & \cdots & 1 \\ 0 & 0 & 0 & \cdots & 0 \\ \vdots & & \vdots & & \vdots \\ 0 & 0 & 0 & \cdots & 0 \end{pmatrix}.$$

That is, if there is a pivot in every column ($Q = n$ and $j_1, \ldots, j_Q = 1, \ldots, n$, respectively) then the reduced row echelon form of A has first n rows equal to the $n \times n$ identity matrix, and remaining rows all zero.

Since Gaussian elimination (which uses only the 3 elementary row operations) does not change the null space we have

$$N(A) = N(\mathrm{rref}\,A),$$

so to find $N(A)$ we instead only need solve the simpler problem of finding $N(\mathrm{rref}\,A)$.

Notice in case $Q = 0$ we have the trivial case $A = O$ and $N(A) = \mathbb{R}^n$ and when $Q = n$ we have a pivot in every column and $N(A) = N(\mathrm{rref}\,A) = \{\underline{0}\}$, so we assume from now on that $Q \in \{1, \ldots, n - 1\}$, and we proceed to compute $N(\mathrm{rref}\,A)$.

If we let b_{ij} denote the element in the i-th row and j-th column of $\mathrm{rref}\,A$, then see that, for $i \in \{1, \ldots, Q\}$, equation number i in the system $\mathrm{rref}\,A\underline{x} = \underline{0}$ is satisfied $\iff x_{j_i} = -\sum_{k \neq j_1, \ldots, j_Q} x_k b_{ik}$, and of course the last $m - Q$ equations are all trivially true for all $\underline{x}$ because the last $m - Q$ rows of $\mathrm{rref}\,A$ are zero. Thus, since for any vector $\underline{x} \in \mathbb{R}^n$ we can write $\underline{x} = \sum_{k=1}^n x_k \underline{e}_k = \sum_{i=1}^Q x_{j_i} \underline{e}_{j_i} + \sum_{k \neq j_1, \ldots, j_Q} x_k \underline{e}_k$, we see that

$$A\underline{x} = \underline{0} \iff \mathrm{rref}\,A\underline{x} = \underline{0} \iff x_{j_i} = -\sum_{k \neq j_1, \ldots, j_Q} x_k b_{ik}, \quad i = 1, \ldots, Q$$
$$\iff \underline{x} = -\sum_{i=1}^Q \left(\sum_{k \neq j_1, \ldots, j_Q} x_k b_{ik}\right) \underline{e}_{j_i} + \sum_{k \neq j_1, \ldots, j_Q} x_k \underline{e}_k$$
$$\iff \underline{x} = \sum_{k \neq j_1, \ldots, j_Q} x_k \left(\underline{e}_k - \sum_{i=1}^Q b_{ik} \underline{e}_{j_i}\right)$$

with x_k arbitrary for $k \neq j_1, \ldots, j_Q$. Thus,

$$N(A) = \text{span}\{\underline{e}_k - \sum_{i=1}^{Q} b_{ik}\underline{e}_{j_i} : k \neq j_1, \ldots, j_Q\},$$

and the $n - Q$ vectors on the right here are clearly l.i., so in fact we have dim $N(A) = n - Q$.

Observe also that rref$A$ has the standard basis vectors $\underline{e}_1, \ldots, \underline{e}_Q$ in columns number $j_1, \ldots, j_Q$, and all the other columns have their last $m - Q$ entries $= 0$, so trivially the column space of rref$A$ is spanned by the l.i. vectors $\underline{e}_1, \ldots, \underline{e}_Q$: if $k \neq j_1, \ldots, j_Q$ then the k-th column β_k of rref$A = \sum_{\ell=1}^{Q} c_\ell \underline{e}_\ell$ for suitable constants $c_1, \ldots, c_Q$. Since the j-th column of a matrix is obtained by multiplying the matrix by the j-th standard basis vector $\underline{e}_j$, in terms of matrix multiplication this says exactly rref$A(\underline{e}_k - \sum_{\ell=1}^{Q} c_\ell \underline{e}_{j_\ell}) = \underline{0}$—i.e., the vector $\underline{e}_k - \sum_{\ell=1}^{Q} c_\ell \underline{e}_{j_\ell}$ is in $N(\text{rref}A)$. Since $N(\text{rref}A) = N(A)$ we must then have that $\underline{e}_k - \sum_{\ell=1}^{Q} c_\ell \underline{e}_{j_\ell}$ is in $N(A)$, or in other words, $A(\underline{e}_k - \sum_{\ell=1}^{Q} c_\ell \underline{e}_{j_\ell}) = \underline{0}$. That is, for each $k \neq j_1, \ldots, j_Q$, the k-th column $\underline{\alpha}_k$ of A is the linear combination $\sum_{\ell=1}^{Q} c_\ell \underline{\alpha}_{j_\ell}$. Thus, we have proved that the column space $C(A)$ of A is spanned by the columns number $j_1, \ldots, j_Q$ of A (i.e., the columns $\underline{\alpha}_{j_1}, \ldots, \underline{\alpha}_{j_Q}$ of A, where $j_1, \ldots, j_Q$ are the numbers of the pivot columns of rref$A$). We also claim that $\underline{\alpha}_{j_1}, \ldots, \underline{\alpha}_{j_Q}$ are l.i., because otherwise there are scalars $c_1, \ldots, c_Q$ not all zero such that $\sum_{\ell=1}^{Q} c_\ell \underline{\alpha}_{j_\ell} = \underline{0}$ and this says exactly that the vector $\sum_{\ell=1}^{Q} c_\ell \underline{e}_{j_\ell} \in N(A) = N(\text{rref}A)$, with $c_1, \ldots, c_Q$ not all zero, so that $\sum_{\ell=1}^{Q} c_\ell \underline{e}_\ell = \text{rref}A(\sum_{\ell=1}^{Q} c_\ell \underline{e}_{j_\ell}) = \underline{0}$ with $c_1, \ldots, c_Q$ not all zero, which is of course not true because $\underline{e}_\ell$, $\ell = 1, \ldots, Q$, are l.i. vectors. Observe that this part of the argument (that $\underline{\alpha}_{j_1}, \ldots, \underline{\alpha}_{j_Q}$ are l.i.) is also correct if $Q = n$; of course in this case we must have $j_1 = 1, j_2 = 2, \ldots, j_n = n$ and the result is that all n columns of A are l.i.

Thus, to summarize, we have proved that if the pivot columns of rref$A$ are column numbers $j_1, \ldots, j_Q$ ($Q \geq 1$), then the columns $\underline{\alpha}_{j_1}, \ldots, \underline{\alpha}_{j_Q}$ are a basis for the column space of A. In particular, Q (the number of nonzero rows in rref$A$) is equal to the dimension of $C(A)$, the column space of A. Since we already proved directly that $\dim(N(A)) = n - Q$, we have thus given a second proof of the rank/nullity theorem and at the same time we have shown how to explicitly find the null space $N(A)$ and the column space $C(A)$.

SECTION 9 EXERCISES

9.1 Write down a single homogeneous linear equation with unknowns x_1, x_2, x_3 such that the solution set consists of the span of the 2 vectors $\begin{pmatrix} 1 \\ 2 \\ -1 \end{pmatrix}, \begin{pmatrix} 1 \\ -2 \\ 0 \end{pmatrix}$.

9.2 Use Gaussian elimination to show that the solution set of the homogeneous system $A\underline{x} = \underline{0}$ with matrix

$$A = \begin{pmatrix} 1 & 2 & 3 & 4 \\ 1 & 4 & 3 & 2 \\ 2 & 5 & 6 & 7 \\ 1 & 0 & 3 & 6 \end{pmatrix}$$

is a plane (i.e., the span of 2 l.i. vectors), and find explicitly 2 l.i. vectors whose span is the solution space.

(b) Find the dimension of the column space of the above matrix (i.e., the dimension of the subspace of $\mathbb{R}^4$ which is spanned by the columns of the matrix).

Hint: Use the result of (a) and the rank nullity theorem.

9.3 Suppose

$$A = \begin{pmatrix} 1 & 0 & 1 & 1 & 1 \\ 2 & 1 & 1 & 3 & 1 \\ 1 & 1 & 2 & 0 & 2 \\ 0 & 0 & 1 & -1 & 1 \end{pmatrix}.$$

Find a basis for the null space $N(A)$ and the column space $C(A)$ of A.

10   INHOMOGENEOUS SYSTEMS

Notice that Gaussian elimination can also be used to solve inhomogeneous systems

$$10.1 \hspace{4cm} A\underline{x} = \underline{b} \, ,$$

where A is $m \times n$, $\underline{b}$ is a given vector in $\mathbb{R}^m$, and $\underline{x}$ is to be determined. First we have the following lemma, which establishes that the set of all solutions of such an inhomogeneous system is either empty or an affine space which is a translate of the null space $N(A)$ of A.

10.2 Lemma. *Suppose A is $m \times n$ and $\underline{b} \in \mathbb{R}^m$. Then:*

(i) $\qquad \exists$ *a solution $\underline{x} \in \mathbb{R}^n$ of* 10.1 $\Longleftrightarrow \underline{b} \in C(A)$ *(the column space of A);*

(ii) $\qquad$ *if $\underline{x}_0 \in \mathbb{R}^n$ is any solution of* 10.1 *then the complete set of solutions*
$$\underline{x} \text{ of } 10.1 \text{ is the affine space } \underline{x}_0 + N(A).$$

10.3 Remark: If V is a subspace of $\mathbb{R}^n$ and $\underline{a} \in \mathbb{R}^n$, then "the affine space through $\underline{a}$ parallel to V" means the translate of V by the vector $\underline{a}$; that is, $\underline{a} + V$ $(= \{\underline{a} + \underline{x} : \underline{x} \in V\})$. With this terminology, part (ii) of the above theorem says that the set of all solutions of 10.1 is the affine space through $\underline{x}_0$ parallel to $N(A)$, assuming that there is at least one solution $\underline{x} = \underline{x}_0$ of 10.1.

Proof of 10.2: Observe first that if $\underline{\alpha}_j$ is the j-th column of A then by definition of matrix multiplication 6.1(ii) we have $A\underline{x} = \sum_{j=1}^n x_j \underline{\alpha}_j$ which means that the set of all vectors $A\underline{x}$ corresponding

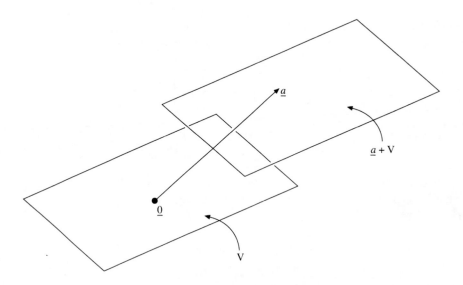

Figure 1.2: Schematic picture of the affine space $\underline{a} + V$ with $n = 3$, dim $V = 2$.

to $\underline{x} \in \mathbb{R}^n$ is just the set of all linear combinations of the columns $\underline{\alpha}_1, \ldots, \underline{\alpha}_n$, i.e., the column space of A. Hence, $A\underline{x} = \underline{b}$ has a solution if and only if $\underline{b} \in C(A)$, so (i) is proved.

Suppose $\underline{x}_0 \in \mathbb{R}^n$ with $A\underline{x}_0 = \underline{b}$ and observe that then

$$A\underline{x} = \underline{b} \iff A(\underline{x} - \underline{x}_0) + A\underline{x}_0 = \underline{b} \iff A(\underline{x} - \underline{x}_0) = \underline{0}$$
$$\iff \underline{x} - \underline{x}_0 \in N(A) \iff \underline{x} \in \underline{x}_0 + N(A),$$

so (ii) is also proved.

In practice, we can check if there is a solution $\underline{x}_0$, and if so actually find the whole solution set, by using Gaussian elimination as follows: We start with the $m \times (n + 1)$ "augmented matrix" $A|\underline{b}$ which has columns $\underline{\alpha}_1, \ldots, \underline{\alpha}_n, \underline{b}$ and use the same elementary row operations which reduce A to rref$A$, but applied to the augmented matrix instead of A: this gives a new $m \times (n + 1)$ augmented matrix rref$A|\tilde{\underline{b}}$ such that rref$A\underline{x} = \tilde{\underline{b}}$ has the same set of solutions as the original system $A\underline{x} = \underline{b}$. One can then directly check whether the new system rref$A\underline{x} = \tilde{\underline{b}}$ has a solution, and, if so, actually find all the solutions $\underline{x}_0 + N(A)$.

SECTION 10 EXERCISES

10.1 Find a cubic polynomial $ax^3 + bx^2 + cx + d$ whose graph passes through the 4 points $(-1, 0)$, $(0, 1)$, $(1, -2)$, $(2, 5)$. Is this cubic polynomial the unique such?

10.2 For each of the inhomogeneous systems with the indicated augmented matrix $A|\underline{b}$: (a) check whether or not a solution exists, and (b) in case a solution exists find the set of all solutions (describe your solution in terms of an affine space $\underline{x}_0 + V$, giving $\underline{x}_0$ and a basis for the subspace V in each case):

(i)

$$A|\underline{b} = \begin{pmatrix} 1 & 0 & 1 & 1 & 1 & | & 1 \\ 2 & 1 & 1 & 3 & 1 & | & 1 \\ 1 & 1 & 2 & 0 & 2 & | & 2 \\ 0 & 0 & 1 & -1 & 1 & | & 1 \end{pmatrix}$$

(ii)

$$A|\underline{b} = \begin{pmatrix} 1 & 0 & 1 & 1 & 2 & | & 1 \\ 2 & 1 & 1 & 3 & 1 & | & 1 \\ 1 & 1 & 2 & 0 & 2 & | & 2 \\ 0 & 0 & 1 & -1 & 1 & | & 1 \end{pmatrix}.$$

CHAPTER 2

Analysis in $\mathbf{R}^n$

1   OPEN AND CLOSED SETS IN $\mathbf{R}^n$

We begin with the definition of open and closed sets in $\mathbb{R}^n$. In these definitions we use the notation that for $\rho > 0$ and $\underline{y} \in \mathbb{R}^n$ the open ball of radius ρ and center $\underline{y}$ (i.e., $\{\underline{x} \in \mathbb{R}^n : \|\underline{x} - \underline{y}\| < \rho\}$) is denoted $B_\rho(\underline{y})$.

1.1 Definition: A subset $U \subset \mathbb{R}^n$ is open if for each $\underline{a} \in U$ there is a $\rho > 0$ such that the ball $B_\rho(\underline{a}) \subset U$.

Observe that direct use of the above definition yields the following general facts:

- $\mathbb{R}^n$ and $\emptyset$ (the empty set) are both open.
- If $U_1, \ldots, U_N$ is any finite collection of open sets, then $\cap_{j=1}^N U_j$ is open (thus "the intersection of finitely many open sets is again open").
- If $\{U_\alpha\}_{\alpha \in \Gamma}$ (Γ any nonempty indexing set) is any collection of open sets, then $\cup_{\alpha \in \Gamma} U_\alpha$ is open (thus "the union of any collection of open sets is again open").

In the definition of closed set we need the notion of *limit point of a set* $C \subset \mathbb{R}^n$: a point $\underline{y} \in \mathbb{R}^n$ is said to be a limit point of a set $C \subset \mathbb{R}^n$ if there is a sequence $\{\underline{x}^{(k)}\}_{k=1,2,\ldots}$ with $\underline{x}^{(k)} \in C$ for each k and $\lim_{k \to \infty} \underline{x}^{(k)} = \underline{y}$ (i.e., $\lim_{k \to \infty} \|x^{(k)} - \underline{y}\| = \underline{0}$). For example, in the case $n = 1$ the point 0 is a limit point of the open interval $(0, 1)$, because $\{\frac{1}{k+1}\}_{k=1,2,\ldots}$ is a sequence of points in $(0, 1)$ with limit zero. Notice that any point $\underline{y} \in C$ is trivially a limit point of C, because it is the limit of the constant sequence $\{\underline{x}^{(k)}\}_{k=1,2,\ldots}$ with $\underline{x}^{(k)} = \underline{y}$ for each k, hence it is always true that

$$C \subset \{\text{limit points of } C\} .$$

1.2 Definition: A subset $C \subset \mathbb{R}^n$ is closed if C contains all its limit points.

There is a very important connection between open and closed sets.

1.3 Lemma. *A set $U \subset \mathbb{R}^n$ is open* $\iff$ $\mathbb{R}^n \setminus U$ *is closed.*

Before we give the proof, we point out an equivalent version of the same result.

1.4 Corollary. *A set $C \subset \mathbb{R}^n$ is closed* $\iff$ $\mathbb{R}^n \setminus C$ *is open.*

Proof of Corollary: Apply 1.3 to the case $U = \mathbb{R}^n \setminus C$; this works because $U = \mathbb{R}^n \setminus C \iff C = \mathbb{R}^n \setminus U$.

Proof of 1.3 $\Rightarrow$: Assume U is open and let $\underline{y}$ be a limit point of $\mathbb{R}^n \setminus U$. If $\underline{y} \in U$ then by definition of open there is $\rho > 0$ such that $B_\rho(\underline{y}) \subset U$, so all points of $\mathbb{R}^n \setminus U$ have distance at least ρ from $\underline{y}$, so, in particular, $\underline{y}$ cannot be a limit point of $\mathbb{R}^n \setminus U$, contradicting the choice of $\underline{y}$. Thus, we have eliminated the possibility that $\underline{y} \in U$ and hence $\underline{y} \in \mathbb{R}^n \setminus U$.

Proof of 1.3 $\Leftarrow$: Take $\underline{y} \in U$. Then $\underline{y}$ is not a limit point of $\mathbb{R}^n \setminus U$ (because $\mathbb{R}^n \setminus U$ contains all of its limit points) and we claim there must be some $\rho > 0$ such that $B_\rho(\underline{y}) \cap \mathbb{R}^n \setminus U = \emptyset$. Indeed, otherwise we would have $B_\rho(\underline{y}) \cap \mathbb{R}^n \setminus U \neq \emptyset$ for every $\rho > 0$, and, in particular, $B_{\frac{1}{k}}(\underline{y}) \cap \mathbb{R}^n \setminus U \neq \emptyset$ for each $k = 1, 2, \ldots$, so for each $k = 1, 2, \ldots$ there is a point $\underline{x}^{(k)} \in B_{\frac{1}{k}}(\underline{y}) \cap \mathbb{R}^n \setminus U$, hence $\|\underline{x}^{(k)} - \underline{y}\| < \frac{1}{k}$ and $\underline{y}$ is a limit point of $\mathbb{R}^n \setminus U$, a contradiction. So in fact there is indeed a $\rho > 0$ such that $B_\rho(\underline{y}) \cap \mathbb{R}^n \setminus U = \emptyset$, or, in other words, $B_\rho(\underline{y}) \subset U$.

Notice that in view of the *De Morgan Laws* (see Exercise 1.5 below) that $\mathbb{R}^n \setminus (\cup_{\alpha \in \Gamma} U_\alpha) = \cap_{\alpha \in \Gamma} (\mathbb{R}^n \setminus U_\alpha)$ and $\mathbb{R}^n \setminus (\cap_{\alpha \in \Gamma} U_\alpha) = \cup_{\alpha \in \Gamma} (\mathbb{R}^n \setminus U_\alpha)$ and the general properties of open sets mentioned above, we conclude the following general properties of closed sets:

- $\mathbb{R}^n$ and $\emptyset$ are both closed.

- If $C_1, \ldots, C_N$ is any finite collection of closed sets, then $\cup_{j=1}^N C_j$ is closed (thus "the union of finitely many closed sets is again closed").

- If $\{C_\alpha\}_{\alpha \in \Gamma}$ (Γ any nonempty indexing set) is any collection of closed sets then $\cap_{\alpha \in \Gamma} C_\alpha$ is closed (thus "the intersection of any collection of closed sets is again closed").

SECTION 1 EXERCISES

1.1 Prove that $\{(x, y) \in \mathbb{R}^2 : y > x^2\}$ is an open set in $\mathbb{R}^2$ which is not closed, and $\{(x, y) \in \mathbb{R}^2 : y \leq x^2\}$ is closed but not open.

1.2 Let $A \subset \mathbb{R}^n$, and define $E \subset A$ to be *relatively open in A* if for each $\underline{x} \in E$ there is $\delta > 0$ such that $B_\delta(\underline{x}) \cap A \subset E$.

Prove that $E \subset A$ is relatively open in A $\Longleftrightarrow$ $\exists$ an open set $U \subset \mathbb{R}^n$ such that $E = A \cap U$.

1.3 Check the claim made above that $U_1, \ldots, U_N$ open $\Rightarrow \cap_{j=1}^N U_j$ is open.

1.4 Give an example, in the case $n = 1$, to show that the result of the previous exercise does not extend to the intersection of infinitely many open sets; thus, give an example of open sets $U_1, U_2, \ldots$ in $\mathbb{R}$ such that $\cap_{j=1}^\infty U_j$ is not open.

1.5 Prove the first De Morgan law mentioned above; that is, if $\{A_\alpha\}_{\alpha \in \Gamma}$ is any collection of subsets of $\mathbb{R}^n$, then $\mathbb{R}^n \setminus (\cup_{\alpha \in \Gamma} A_\alpha) = \cap_{\alpha \in \Gamma} (\mathbb{R}^n \setminus A_\alpha)$.

Hint: You have to prove $\mathbb{R}^n \setminus (\cup_{\alpha \in \Gamma} A_\alpha) \subset \cap_{\alpha \in \Gamma} (\mathbb{R}^n \setminus A_\alpha)$ and $\mathbb{R}^n \setminus (\cup_{\alpha \in \Gamma} A_\alpha) \supset \cap_{\alpha \in \Gamma} (\mathbb{R}^n \setminus A_\alpha)$, i.e., $\underline{x} \in \mathbb{R}^n \setminus (\cup_{\alpha \in \Gamma} A_\alpha) \Rightarrow \underline{x} \in \cap_{\alpha \in \Gamma} (\mathbb{R}^n \setminus A_\alpha)$ and $\underline{x} \in \cap_{\alpha \in \Gamma} (\mathbb{R}^n \setminus A_\alpha) \Rightarrow \underline{x} \in \mathbb{R}^n \setminus (\cup_{\alpha \in \Gamma} A_\alpha)$.

2   BOLZANO-WEIERSTRASS, LIMITS AND CONTINUITY IN R^n

Here we shall write points in $\mathbb{R}^n$ as rows: $\underline{x} = (x_1, \ldots, x_n)$.

We first give the definition of convergence of a sequence $\{\underline{x}^{(k)}\}_{k=1,2,\ldots}$ of such points:

2.1 Definition: Given a sequence $\{\underline{x}^{(k)}\}_{k=1,2,\ldots}$ of points in $\mathbb{R}^n$, $\lim_{k\to\infty} \underline{x}^{(k)} = \underline{y}$ means $\lim_{k\to\infty} \|\underline{x}^{(k)} - \underline{y}\| = 0$.

Notice that this makes sense, because $\{\|\underline{x}^{(k)} - \underline{y}\|\}_{k=1,2,\ldots}$ is a sequence of real numbers, and we are already familiar with the definition of convergence of such real sequences. In terms of ε, N, the Def. 2.1 actually says that $\lim \underline{x}^{(k)} = \underline{y} \iff$ for each $\varepsilon > 0$ there is N such that $\|\underline{x}^{(k)} - \underline{y}\| < \varepsilon$ whenever $k \geq N$. In view of the inequalities

2.2
$$\max\{|x_j^{(k)} - y_j| : j = 1, \ldots, n\} \leq \|\underline{x}^{(k)} - \underline{y}\| \leq \sum_{j=1}^{n} |x_j^{(k)} - y_j|$$

it is very easy to check the following lemma (see Exercise 2.2 below).

2.3 Lemma. $\lim x^{(k)} = \underline{y}$ *with* $\underline{y} = (y_1, \ldots, y_n) \iff \lim x_j^{(k)} = y_j \, \forall \, j = 1, \ldots, n$.

2.4 Remark: The real sequences $\{x_j^{(k)}\}_{k=1,2,\ldots}$ (for $j = 1, \ldots, n$) are referred to as *the component sequences* corresponding to the vector sequence $\underline{x}^{(k)}$, so with this terminology Lem. 2.3 says that the vector sequence $\underline{x}^{(k)}$ converges if and only if each of the component sequences converge, and in this case the limit of the vector sequence is the point in $\mathbb{R}^n$ with j-th component equal to the limit of the j-th component sequence, $j = 1, \ldots, n$.

Recall from Lecture 2 of the Appendix that every bounded sequence in $\mathbb{R}$ has a convergent subsequence. A similar theorem is true for sequences in $\mathbb{R}^n$.

2.5 Theorem (Bolzano-Weierstrass in R^n.) *If* $\{\underline{x}^{(k)}\}_{k=1,2,\ldots}$ *is a bounded sequence in $\mathbb{R}^n$ (so that there is a constant R with $\|\underline{x}^{(k)}\| \leq R \, \forall \, k = 1, 2, \ldots$), then there is a convergent subsequence $\{\underline{x}^{(k_j)}\}_{j=1,2,\ldots}$.*

Proof: As we mentioned above, this is already proved in Lecture 2 of the Appendix for the case $n = 1$, and the general case follows from this and induction on n. The details are left as an exercise (see Exercise 2.3 below).

Now suppose A is an arbitrary subset of $\mathbb{R}^n$, and let $f : A \to \mathbb{R}^m$, where $m \geq 1$. We want to define continuity of f at a point $\underline{a} \in A$. The definition which follows is similar to the the special case discussed in Lecture 3 of the Appendix A—i.e. the special case when A is an interval in $\mathbb{R}$ and $m = 1$. In the general case when A is any subset of $\mathbb{R}^n$ and $m \geq 1$ the definition is in fact as follows:

Definition (i): $f : A \to \mathbb{R}^m$ is continuous at the point $\underline{a} \in A$ if for each $\varepsilon > 0$ there is $\delta > 0$ such that $\underline{x} \in B_\delta(\underline{a}) \cap A \Rightarrow \|f(\underline{x}) - f(\underline{a})\| < \varepsilon$.

(ii) $f : A \to \mathbb{R}^m$ is said to be continuous if is continuous at every point $\underline{a} \in A$.

Analogous to the 1-variable theorem discussed in Lecture 3 of Appendix A, we have the following theorem in $\mathbb{R}^n$.

2.6 Theorem. *Suppose $K \subset \mathbb{R}^n$ is compact (i.e., K is a closed bounded subset of $\mathbb{R}^n$) and $f : K \to \mathbb{R}$ is continuous. Then f attains its maximum and minimum values on K; that is, there are points $\underline{a}, \underline{b} \in K$ with $f(\underline{a}) \leq f(\underline{x}) \leq f(\underline{b}) \, \forall \underline{x} \in K$.*

Proof: The proof is almost identical to the proof of the corresponding 1-variable theorem given in Lecture 3 of the Appendix, except that we use the Bolzano-Weierstrass theorem in $\mathbb{R}^n$ rather than in $\mathbb{R}$. (See Exercise 2.4 below.)

We shall also need the concept of of limit of a function which is defined on a subset of $\mathbb{R}^n$. So suppose $f : A \to \mathbb{R}^m$, where $A \subset \mathbb{R}^n$ and $m \geq 1$. We shall define the notion of limit at any point $\underline{a} \in A$ which is not an *isolated point of A*—a point $\underline{a} \in A$ is called an isolated point of A if there is some $\delta_0 > 0$ such that $A \cap B_{\delta_0}(\underline{a}) = \{\underline{a}\}$ (i.e., if there is some ball $B_{\delta_0}(\underline{a})$ such that $\underline{a}$ is the only point of A which lies in the ball $B_{\delta_0}(\underline{a})$). Thus $\underline{a} \in A$ is *not* an isolated point of A if and only if $A \cap (B_\delta(\underline{a}) \setminus \{\underline{a}\}) \neq \emptyset$ for each $\delta > 0$.

2.7 Definition: Suppose $A \subset \mathbb{R}^n$, $f : A \to \mathbb{R}^m$, $\underline{a} \in A$, $\underline{a}$ not an isolated point of A, and $\underline{y} \in \mathbb{R}^m$. Then $\lim_{\underline{x} \to \underline{a}} f(\underline{x}) = \underline{y}$ means that for each $\varepsilon > 0$ there is a $\delta > 0$ such that $\underline{x} \in A \cap B_\delta(\underline{a}) \setminus \{\underline{a}\} \Rightarrow \|f(\underline{x}) - \underline{y}\| < \varepsilon$.

2.8 Remark: With the notation of the above definition, notice that $\lim_{\underline{x} \to \underline{a}} f(\underline{x}) = \underline{y}$ with $\underline{y} = (y_1, \ldots, y_m) \iff \lim_{\underline{x} \to \underline{a}} f_j(\underline{x}) = y_j \, \forall j = 1, \ldots, m$.

We leave the proof as an exercise (see Exercise 2.4 below), and we also leave the proofs of the standard limit theorems in the following lemma as an exercise (see Exercise 2.4).

2.9 Lemma. *Suppose $\lambda, \mu \in \mathbb{R}$, $f, g : A \to \mathbb{R}^m$, $\underline{a} \in A$, and $\lim_{\underline{x} \to \underline{a}} f(\underline{x})$ and $\lim_{\underline{x} \to \underline{a}} g(\underline{x})$ both exist. Then:*

(i) $\quad \lim\limits_{\underline{x} \to \underline{a}} (\lambda f(\underline{x}) + \mu g(\underline{x}))$ *exists and* $= \lambda \lim\limits_{\underline{x} \to \underline{a}} f(\underline{x}) + \mu \lim\limits_{\underline{x} \to \underline{a}} g(\underline{x})$

(ii) $\quad m = 1 \Rightarrow \lim\limits_{\underline{x} \to \underline{a}} (f(\underline{x})g(\underline{x}))$ *exists and* $= (\lim\limits_{\underline{x} \to \underline{a}} f(\underline{x}))(\lim\limits_{\underline{x} \to \underline{a}} g(\underline{x}))$

(iii) $\quad m = 1$ *and* $\lim\limits_{\underline{x} \to \underline{a}} g(\underline{x}) \neq 0 \Rightarrow \lim\limits_{\underline{x} \to \underline{a}} (f(\underline{x})/g(\underline{x}))$

$\qquad\qquad\qquad\qquad\qquad$ *exists and* $= (\lim\limits_{\underline{x} \to \underline{a}} f(\underline{x}))/(\lim\limits_{\underline{x} \to \underline{a}} g(\underline{x}))$.

2.10 Remark: Suppose $f : A \to \mathbb{R}^m$ and $\underline{a} \in A$. Observe that f is trivially continuous at $\underline{a}$ if $\underline{a}$ is an isolated point of A, and if $\underline{a}$ is not an isolated point of A then f is continuous at $\underline{a}$ if and only if $\lim_{\underline{x} \to \underline{a}} f(\underline{x}) = f(\underline{a})$ (meaning $\lim_{\underline{x} \to \underline{a}} f(\underline{x})$ exists and is equal to $f(\underline{a})$).

SECTION 2 EXERCISES

2.1 (Sandwich theorem.) Suppose $A \subset \mathbb{R}^n, \underline{a} \in A, \underline{y} \in \mathbb{R}^m, f, g, h : A \to \mathbb{R}^m$ and $\lim_{\underline{x} \to \underline{a}} g(\underline{x}) = \lim_{\underline{x} \to \underline{a}} h(\underline{x}) = \underline{y}$. Prove $g(\underline{x}) \leq f(\underline{x}) \leq h(\underline{x}) \, \forall \, \underline{x} \in A \Rightarrow \lim_{\underline{x} \to \underline{a}} f(\underline{x}) = \underline{y}$.

Note: Part of what you have to prove is that $\lim_{\underline{x} \to \underline{a}} f(\underline{x})$ exists.

2.2 Give the proof of 2.2 above.

2.3 Using the Bolzano-Weierstrass theorem for real sequences (from Lecture 2 of Appendix A), give the proof of Thm. 2.5 above.

Caution: An application of the theorem in the case $n = 1$ (Lecture 2 of Appendix A) tells us that each component sequence $\{x_j^{(k)}\}_{k=1,2,\ldots}$ has a convergent subsequence, but you need to take account of the fact that there may be a different subsequence corresponding to each j.

2.4 Give the proofs of Thm. 2.6, Rem. 2.8, and Lem. 2.9 above.

2.5 Suppose K is a compact (i.e., closed and bounded) subset of $\mathbb{R}^n$ and let $f : K \to \mathbb{R}$ be continuous. Prove that for each $\varepsilon > 0$ there is $\delta > 0$ such that $\underline{x}, \underline{y} \in K$ and $\|\underline{x} - \underline{y}\| < \delta \Rightarrow |f(\underline{x}) - f(\underline{y})| < \varepsilon$. Hint: Otherwise there is $\varepsilon = \varepsilon_0 > 0$ such that this fails for each $\delta > 0$, which means it fails for $\delta = \frac{1}{k}$ for each $k = 1, 2, \ldots$, which means that for each k there are points $\underline{x}_k, \underline{y}_k \in K$ with $\|\underline{x}_k - \underline{y}_k\| < \frac{1}{k}$ and $|f(\underline{x}_k) - f(\underline{y}_k)| \geq \varepsilon_0$.

3   DIFFERENTIABILITY IN $\mathbb{R}^n$

Recall first the definition of differentiability of a function $f : (\alpha, \beta) \to \mathbb{R}$ of one variable: f is differentiable at a point $a \in (\alpha, \beta)$ if

3.1
$$\lim_{h \to 0} \frac{f(a + h) - f(a)}{h} \text{ exists .}$$

In this case, the limit is denoted $f'(a)$ and is referred to as the derivative of f at a.

This definition does not extend well to functions of more than 1-variable because in the case of functions $f : U \to \mathbb{R}$ with $U \subset \mathbb{R}^n$ open and $n \geq 2$, difference quotients like $\frac{f(a+h)-f(a)}{h}$ make no sense (we cannot divide by the vector quantity $\underline{h}$). We therefore need an alternative definition of differentiability which makes sense for such functions and which is equivalent to the usual definition 3.1 of differentiability when $n = 1$. The key observation in this regard is that 3.1 (in the case $n = 1$ when $f : (\alpha, \beta) \to \mathbb{R}$ and $a \in (\alpha, \beta)$) is equivalent to the requirement that there is a real number A such that

3.2
$$\lim_{h \to 0} |h|^{-1} |f(a + h) - (f(a) + Ah)| = 0 .$$

Notice that such A exists if and only if $A = \lim \frac{f(a+h)-f(a)}{h}$, because

$$|h|^{-1}|f(a+h) - (f(a) + Ah)| = \left|\frac{f(a+h) - f(a)}{h} - A\right|$$

and hence

$$\lim_{h\to 0} |h|^{-1}|f(a+h) - (f(a) + Ah)| = 0 \iff \lim_{h\to 0} \frac{f(a+h) - f(a)}{h} = A .$$

So 3.2 is equivalent to 3.1 in case $n = 1$, but the point now is that 3.2 has a natural generalization to the case when $f : U \to \mathbb{R}^m$, where $U \subset \mathbb{R}^n$ is open and m, n are arbitrary positive integers, as follows.

3.3 Definition: If $f : U \to \mathbb{R}^m$ then we say that f is differentiable at $\underline{a} \in U$ if there is an $m \times n$ matrix A such that

$$\lim_{\underline{h}\to\underline{0}} \|\underline{h}\|^{-1}\|f(\underline{a} + \underline{h}) - (f(\underline{a}) + A\underline{h})\| = 0 .$$

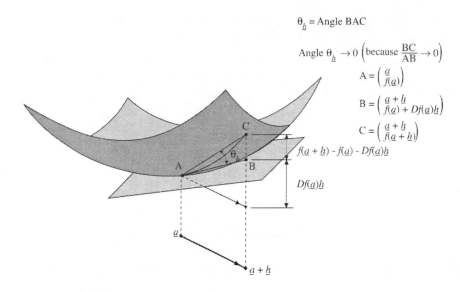

$\theta_{\underline{h}} = $ Angle BAC

Angle $\theta_{\underline{h}} \to 0 \left(\text{because } \frac{BC}{AB} \to 0\right)$

$A = \begin{pmatrix} \underline{a} \\ f(\underline{a}) \end{pmatrix}$

$B = \begin{pmatrix} \underline{a} + \underline{h} \\ f(\underline{a}) + Df(\underline{a})\underline{h} \end{pmatrix}$

$C = \begin{pmatrix} \underline{a} + \underline{h} \\ f(\underline{a} + \underline{h}) \end{pmatrix}$

$f(\underline{a} + \underline{h}) - f(\underline{a}) - Df(\underline{a})\underline{h}$

$Df(\underline{a})\underline{h}$

$\underline{a}$

$\underline{a} + \underline{h}$

Figure 2.1: Definition of differentiability, $n = 2, m = 1$.

In case 3.3 holds, A is called the *derivative matrix (or Jacobian matrix) of the function f at the point $\underline{a}$.* It is unique if it exists at all, and we could fairly easily check that now, but in any case it will be a by-product of the result of Thm. 4.3 in the next section, so we'll leave the proof of uniqueness until then.

The derivative matrix A of f at $\underline{a}$, when it exists, is usually denoted $Df(\underline{a})$.

As for functions of 1-variable we have the fact that differentiability $\Rightarrow$ continuity. More precisely:

3.4 Lemma. *Suppose $f : U \to \mathbb{R}^m$ with U open in $\mathbb{R}^n$ and suppose $\underline{a} \in U$. Then differentiability of f at $\underline{a} \Rightarrow$ continuity of f at $\underline{a}$.*

Proof: Assume f is differentiable at $\underline{a} \in U$. Pick $\rho > 0$ such that $B_\rho(\underline{a}) \subset U$, which we can do because U is open. Then for $0 < \|\underline{h}\| < \rho$ we have

$$
\begin{aligned}
\| f(\underline{a} + \underline{h}) - f(\underline{a}) \| &= \| f(\underline{a} + \underline{h}) - (f(\underline{a}) + Df(\underline{a})\underline{h}) - Df(\underline{a})\underline{h} \| \\
&\leq \| f(\underline{a} + \underline{h}) - (f(\underline{a}) + Df(\underline{a})\underline{h}) \| + \| Df(\underline{a})\underline{h} \| \\
&\leq (\|\underline{h}\|^{-1} \| f(\underline{a} + \underline{h}) - (f(\underline{a}) + Df(\underline{a})\underline{h}) \|) \|\underline{h}\| + \| Df(\underline{a}) \| \|\underline{h}\|
\end{aligned}
$$

and $\lim_{\underline{h} \to \underline{0}} \|\underline{h}\|^{-1}(\| f(\underline{a} + \underline{h}) - (f(\underline{a}) + Df(\underline{a})\underline{h}))\|$, $\lim_{\underline{h} \to \underline{0}} \|\underline{h}\|$, $\lim_{\underline{h} \to \underline{0}} \| Df(\underline{a}) \| \|\underline{h}\|$ are all zero, so, by Lem. 2.9, $\lim_{\underline{h} \to \underline{0}} \| f(\underline{a} + \underline{h}) - f(\underline{a}) \| = 0.0 + 0 = 0$, which is the definition of continuity of f at $\underline{a}$.

4   DIRECTIONAL DERIVATIVES, PARTIAL DERIVATIVES, AND GRADIENT

Notice that if $f : U \to \mathbb{R}^m$ with U open and if $\underline{a} \in U$, we can find $\rho > 0$ such that $B_\rho(\underline{a}) \subset U$ and so, for any vector $\underline{v} \in \mathbb{R}^n$, $f(\underline{a} + t\underline{v})$ is a function of the real variable t, and is well defined for $|t| < \|\underline{v}\|^{-1}\rho$ if $\underline{v} \neq \underline{0}$ and is a constant function $f(\underline{a} + t\underline{v}) \equiv f(\underline{a})$ for all $t \in \mathbb{R}$ if $\underline{v} = \underline{0}$.

4.1 Definition: The *directional derivative of f at $\underline{a}$ by the vector $\underline{v}$*, denoted $D_{\underline{v}} f(\underline{a})$, is defined by

$$
D_{\underline{v}} f(\underline{a}) = \lim_{h \to 0} h^{-1}(f(\underline{a} + h\underline{v}) - f(\underline{a}))
$$

whenever the limit on the right exists. Observe that, since $g(t) = f(\underline{a} + t\underline{v})$ is a function of the real variable t for t in some open interval containing 0, the Def. 4.1 just says that the directional derivative $D_{\underline{v}} f(\underline{a})$ is exactly the derivative $g'(0)$ at $t = 0$ of the function $g(t)$ defined by $g(t) = f(\underline{a} + t\underline{v})$.

4.2 Definition: In the special case $\underline{v} = \underline{e}_j$ (the j-th standard basis vector in $\mathbb{R}^n$) the directional derivative $D_{\underline{e}_j} f(\underline{a})$ is usually denoted $D_j f(\underline{a})$ whenever it exists; thus

$$
D_j f(\underline{a}) = \lim_{h \to 0} h^{-1} f(\underline{a} + h\underline{e}_j - f(\underline{a})) \, .
$$

$D_j f(\underline{a})$ so defined is called the j-th partial derivative of f at $\underline{a}$. An alternative notation (sometimes referred to as "classical notation") is to write

$$
\frac{\partial f(\underline{x})}{\partial x_j} = D_j f(\underline{x}) \, .
$$

Notice that the simplest way of calculating $D_j f(\underline{x})$ is usually simply to take the derivative (as in 1-variable calculus) of the function $f(x_1, \ldots, x_n)$ with respect to the single variable x_j while holding all the other variables, $x_i, i \neq j$, fixed. Thus, in general, it is not necessary to use the formal limit definition of 4.2.

There is an important connection between partial derivatives and the derivative matrix at points where f is differentiable, given in the following theorem.

4.3 Theorem. *If $f : U \to \mathbb{R}^m$ with U open and if f is differentiable at a point $\underline{a} \in U$, then all the partial derivatives of f at $\underline{a}$ exist, and in fact the j-th column of the derivative matrix $Df(\underline{a})$ is exactly the j-th partial derivative $D_j f(\underline{a})$, $j = 1, \ldots, n$. Furthermore, if f is differentiable at $\underline{a}$ then all the directional derivatives $D_{\underline{v}} f(\underline{a})$ exist, and are given by the formula*

$$D_{\underline{v}} f(\underline{a}) = Df(\underline{a})\underline{v},$$

or equivalently,

$$D_{\underline{v}} f(\underline{a}) = \sum_{j=1}^{n} v_j D_j f(\underline{a}).$$

4.4 Caution: The above says, in particular, that f differentiable at $\underline{a}$ implies that all the partial derivatives $D_j f(\underline{a})$, $j = 1, \ldots, n$, and all the directional derivatives $D_{\underline{v}} f(\underline{a})$ exist, but the converse is *false*; that is, it may happen that all the directional derivatives (including the partial derivatives) exist yet f still fails to be differentiable at $\underline{a}$. (See Exercise 4.1 below.)

Proof of Theorem 4.3: Take $\rho > 0$ such that $B_\rho(\underline{a}) \subset U$ and let $A = Df(\underline{a})$. We are given $\lim_{\underline{h} \to \underline{0}} \|\underline{h}\|^{-1} \|f(\underline{a} + \underline{h}) - (f(\underline{a}) + A\underline{h})\| = 0$ so given $\varepsilon > 0$ there is a $\delta \in (0, \rho)$ such that $0 < \|\underline{h}\| < \delta \Rightarrow \|\underline{h}\|^{-1} \|f(\underline{a} + \underline{h}) - (f(\underline{a}) + A\underline{h})\| < \varepsilon$ and taking, in particular, $\underline{h} = h\underline{e}_j$ we see that $h \in \mathbb{R}$ with $0 < |h| < \delta \Rightarrow |h|^{-1} \|f(\underline{a} + h\underline{e}_j) - (f(\underline{a}) + hA\underline{e}_j)\| < \varepsilon \iff \|h^{-1}(f(\underline{a} + h\underline{e}_j) - f(\underline{a})) - A\underline{e}_j\| < \varepsilon$, which is precisely the ε, δ definition of $\lim h^{-1}(f(\underline{a} + h\underline{e}_j) - f(\underline{a})) = A\underline{e}_j$, so in fact $D_j f(\underline{a})$ exists and equals the j-th column of $Df(\underline{a})$. If $\underline{v}$ is any vector in $\mathbb{R}^n \setminus \{\underline{0}\}$ and if $h \in \mathbb{R}$ with $0 < |h| < \delta/\|\underline{v}\|$ then $|h|^{-1}\|f(\underline{a} + h\underline{v}) - (f(\underline{a}) + hA\underline{v})\| < \varepsilon$, and hence $\|h^{-1}(f(\underline{a} + h\underline{v}) - f(\underline{a})) - A\underline{v}\| < \varepsilon$ which, since $\varepsilon > 0$ is arbitrary, says $\lim_{h \to 0} h^{-1}(f(\underline{a} + h\underline{v}) - f(\underline{a})) = A\underline{v}$; that is $D_{\underline{v}} f(\underline{a})$ exists and is equal to $A\underline{v} = Df(\underline{a})\underline{v} = \sum_{j=1}^{n} v_j D_j f(\underline{a})$, since we already proved the j-th column of $Df(\underline{a})$ is $D_j f(\underline{a})$.

4.5 Definition: Let $f : U \to \mathbb{R}$ with U open. The gradient of f at a point $\underline{a} \in U$ is the vector $(D_1 f(\underline{a}), \ldots, D_n f(\underline{a}))^{\mathsf{T}}$.

We claim that the gradient has geometric significance: If f is differentiable at $\underline{a}$ it gives the direction of the fastest rate of change of f when we start from the point $\underline{a}$, and furthermore it's magnitude $\|\nabla f(\underline{a})\|$ gives the actual rate of change. More precisely:

4.6 Lemma. *Suppose $f : U \to \mathbb{R}$ with U open, let $\underline{a} \in U$ and assume that f is differentiable at $\underline{a}$. Then $\max\{D_{\underline{v}} f(\underline{a}) : \underline{v} \in \mathbb{R}^n, \ \|\underline{v}\| = 1\} = \|\nabla f(\underline{a})\|$, and for $\nabla f(\underline{a}) \neq \underline{0}$ the maximum is attained when and only when $\underline{v} = \|\nabla f(\underline{a})\|^{-1} \nabla f(\underline{a})$.*

Proof: By Lem. 4.4 we have $D_{\underline{v}} f(\underline{a}) = \sum_{j=1}^{n} v_j D_j f(\underline{a}) = \underline{v} \cdot \nabla f(\underline{a})$, and by the version of the Cauchy-Schwarz inequality in Ch. 1, Exercise 2.2 we have for $\|\underline{v}\| = 1$ that

(1) $$\underline{v} \cdot \nabla f(\underline{a}) \leq \|\nabla f(\underline{a})\|$$

and equality holds in this inequality if and only if $\nabla f(\underline{a}) = \lambda \underline{v}$ with $\lambda \geq 0$. But $\nabla f(\underline{a}) = \lambda \underline{v}$ with $\lambda \geq 0$ and $\|\underline{v}\| = 1 \Rightarrow \lambda = \|\nabla f(\underline{a})\|$, so if $\nabla f(\underline{a}) \neq \underline{0}$ equality holds in (1) if and only if $\underline{v} = \|\nabla f(\underline{a})\|^{-1} \nabla f(\underline{a})$, as claimed.

In view of the remarks in 4.4 above, we have the inconvenient fact that all the partial derivatives of f might exist, yet f still not be differentiable. Fortunately, there is after all a criterion, involving only the partial derivatives of f, which guarantees differentiability of f, as follows.

4.7 Theorem. *Let $f : U \to \mathbb{R}^m$ and assume that there is a ball $B_\rho(\underline{a}) \subset U$ such that the partial derivatives $D_j f(\underline{x})$ exist at each point $\underline{x} \in B_\rho(\underline{a})$ and are continuous at $\underline{a}$. Then f is differentiable at $\underline{a}$.*

4.8 Remark: In particular, if the partial derivatives $D_j f$ exist everywhere in U and are continuous at each point of U (in this case we say that "f is a C^1 function on U"), then we can apply the above theorem at every point of U, so f is differentiable at every point of U in this case.

Proof of Theorem 4.7: Observe that it suffices to prove the theorem in the case $m = 1$, because in case $m \geq 2$ we have $f(\underline{x}) = (f_1(\underline{x}), \ldots, f_m(\underline{x}))^{\mathrm{T}}$ and we can apply the case $m = 1$ to each individual component f_j, which evidently implies the required result. With $m = 1$ we have f real-valued and for $0 < |\underline{h}| < \rho$ we have

$$(1) \qquad f(\underline{a} + \underline{h}) - f(\underline{a}) = \sum_{j=1}^{n} \left(f(\underline{a} + \underline{h}^{(j)}) - f(\underline{a} + \underline{h}^{(j-1)}) \right),$$

where we use the notation $\underline{h}^{(j)} = \sum_{i=1}^{j} h_i \underline{e}_i$ for $j = 1, \ldots, n$ and $\underline{h}^{(0)} = \underline{0}$. Observe that then $\underline{h}^{(j)} = \underline{h}^{(j-1)} + h_j \underline{e}_j$ for $j = 1, \ldots, n$, and so in the difference $f(\underline{a} + \underline{h}^{(j)}) - f(\underline{a} + \underline{h}^{(j-1)}) = f(\underline{a} + \underline{h}^{(j-1)} + h_j \underline{e}_j) - f(\underline{a} + \underline{h}^{(j-1)})$ we are varying only the j-th variable (i.e., a single variable), so the Mean Value Theorem from 1-variable calculus implies $f(\underline{a} + \underline{h}^{(j)}) - f(\underline{a} + \underline{h}^{(j-1)}) = h_j D_j f(\underline{h}^{(j-1)} + \theta_j h_j \underline{e}_j)$ for some $\theta_j \in (0, 1)$, whence (1) implies

$$f(\underline{a} + \underline{h}) - f(\underline{a}) = \sum_{j=1}^{n} h_j D_j f(\underline{a} + \underline{h}^{(j-1)} + \theta_j h_j \underline{e}_j),$$

and hence

$$f(\underline{a} + \underline{h}) - f(\underline{a}) - Df(\underline{a})\underline{h} = \sum_{j=1}^{n} h_j \left(D_j f(\underline{a} + \underline{h}^{(j-1)} + \theta_j h_j \underline{e}_j) - D_j f(\underline{a}) \right).$$

and in particular, with $Df = (D_1 f, \ldots, D_n f)$,

$$
(2) \quad \|f(\underline{a} + \underline{h}) - f(\underline{a}) - Df(\underline{a})\underline{h}\| = \left\| \sum_{j=1}^{n} h_j \left(D_j f(\underline{a} + \underline{h}^{(j-1)} + \theta_j h_j \underline{e}_j) - D_j f(\underline{a}) \right) \right\|
$$

$$
\leq \sum_{j=1}^{n} |h_j| \left\| D_j f(\underline{a} + \underline{h}^{(j-1)} + \theta_j h_j \underline{e}_j) - D_j f(\underline{a}) \right\|
$$

$$
\leq \|\underline{h}\| \sum_{j=1}^{n} \left\| D_j f(\underline{a} + \underline{h}^{(j-1)} + \theta_j h_j \underline{e}_j) - D_j f(\underline{a}) \right\| ,
$$

where we used the triangle inequality (2.6 of Ch. 1) to go from the first line to the second.

Let $\varepsilon > 0$. Since $D_j f(\underline{x})$ is continuous at $\underline{x} = \underline{a}$, for each $j = 1, \ldots, n$ we have $\delta_j \in (0, \rho)$ such that

$$
\|\underline{k}\| < \delta_j \Rightarrow \|D_j f(\underline{a} + \underline{k}) - D_j f(\underline{a})\| < \varepsilon/n ,
$$

and hence with $\delta = \min\{\delta_1, \ldots, \delta_n\}$ we have

$$
(3) \quad \|\underline{k}\| < \delta \Rightarrow \|D_j f(\underline{a} + \underline{k}) - D_j f(\underline{a})\| < \varepsilon/n \text{ for each } j = 1, \ldots, n .
$$

Finally, observe that if $\|\underline{h}\| < \delta$ then $\|\underline{h}^{(j-1)} + \theta_j h_j \underline{e}_j\| = \sqrt{\sum_{i=1}^{j-1} h_i^2 + \theta_j^2 h_j^2} \leq \|\underline{h}\| < \delta$, and hence we can use (3) with $\underline{k} = \underline{h}^{(j-1)} + \theta_j h_j \underline{e}_j$ and so (2) implies

$$
\|f(\underline{a} + \underline{h}) - f(\underline{a}) - Df(\underline{a})\underline{h}\| < \varepsilon \|\underline{h}\| \text{ whenever } 0 < \|\underline{h}\| < \delta ,
$$

thus proving that f is differentiable at $\underline{a}$.

SECTION 4 EXERCISES

4.1 Let f be the function on $\mathbb{R}^2$ defined by $f(x, y) = \begin{cases} \frac{x^3}{y} & \text{if } y \neq 0 \\ 0 & \text{if } y = 0. \end{cases}$

(i) Prove that the directional derivative $D_{\underline{v}} f(0, 0)$ (exists and) $= 0$ for each $\underline{v} \in \mathbb{R}^2$.

(ii) f is not continuous at $(0, 0)$.

(iii) f is not differentiable at $(0, 0)$.

4.2 For the function $f(x_1, x_2) = \begin{cases} \frac{|x_1||x_2|}{\sqrt{x_1^2 + x_2^2}} & \text{if } (x_1, x_2) \neq (0, 0) \\ 0 & \text{if } (x_1, x_2) = (0, 0), \end{cases}$

(i) Find the directional derivatives at $(0, 0)$ (and show they all exist),

(ii) Show that the formula $D_{\underline{v}} f(0, 0) = \sum_{j=1}^{2} v_j D_j f(0, 0)$ fails for some vectors $\underline{v} \in \mathbb{R}^2$,

(iii) Hence, using (ii) together with the appropriate theorem, prove that f is not differentiable at $(0, 0)$.

5   CHAIN RULE

The chain rule for functions of one variable says that the composite function $g \circ f$ is differentiable at a point x with derivative $g'(f(x))f'(x)$ provided both the derivatives $f'(x)$ and $g'(f(x))$ exist. The chain rule for functions of more than one variable is similar, and the proof is more or less identical, except that the derivatives f', g' must be replaced by the relevant derivative matrices. The precise theorem is as follows.

5.1 Theorem (Chain Rule.) *Suppose $U \subset \mathbb{R}^n$ and $V \subset \mathbb{R}^m$ are open and $f : U \to V$, $g : V \to \mathbb{R}^\ell$ are given, and suppose that f is differentiable at a point $\underline{x} \in U$ and that g is differentiable at the point $\underline{y} = f(\underline{x}) \in V$. Then $g \circ f$ is differentiable at $\underline{x}$ and $D(g \circ f)(\underline{x}) = Dg(f(\underline{x}))Df(\underline{x})$.*

5.2 Remark: $Dg(f(\underline{x}))Df(\underline{x})$ is the product of the $\ell \times m$ matrix $Dg(f(\underline{x}))$ times the $m \times n$ matrix $Df(\underline{x})$, which makes sense and gives an $\ell \times n$ matrix, which is at least the correct size matrix to possibly represent $D(g \circ f)(\underline{x})$, because the composite $g \circ f$ maps the set $U \subset \mathbb{R}^n$ into $\mathbb{R}^\ell$, so at least the statement of the theorem makes sense.

Proof of the Chain Rule: Let $\varepsilon \in (0, 1)$ and take $\delta_0 > 0$ such that $B_{\delta_0}(\underline{x}) \subset U$ and $B_{\delta_0}(f(\underline{x})) \subset V$. Observe first that by differentiability of g at $\underline{y} = f(\underline{x})$ we can choose $\delta_1 \in (0, \delta_0)$ such that

$$\|\underline{z} - \underline{y}\| < \delta_1 \Rightarrow \|g(\underline{z}) - g(\underline{y}) - Dg(\underline{y})(\underline{z} - \underline{y})\| \leq \varepsilon\|\underline{z} - \underline{y}\|$$

and, on the other hand, by differentiability of f at $\underline{x}$ we can choose $\delta_2 \in (0, \delta_0)$ such that

$$\|\underline{h}\| < \delta_2 \Rightarrow \|f(\underline{x} + \underline{h}) - f(\underline{x}) - Df(\underline{x})\underline{h}\| \leq \varepsilon\|\underline{h}\|$$

and, in particular,

$$\|\underline{h}\| < \delta_2 \Rightarrow \|f(\underline{x} + \underline{h}) - f(\underline{x})\| = \|f(\underline{x} + \underline{h}) - f(\underline{x}) - Df(\underline{x})\underline{h} + Df(\underline{x})\underline{h}\|$$
$$\leq \|f(\underline{x} + \underline{h}) - f(\underline{x}) - Df(\underline{x})\underline{h}\| + \|Df(\underline{x})\underline{h}\|$$
$$\leq \varepsilon\|\underline{h}\| + \|Df(\underline{x})\|)\|\underline{h}\|$$
$$\leq (1 + \|Df(\underline{x})\|)\|\underline{h}\| < \delta_1$$

if $\|\underline{h}\| < (1 + \|Df(\underline{x})\|)^{-1}\delta_1$. So, in particular, if $\|\underline{h}\| < \min\{(1 + \|Df(\underline{x})\|)^{-1}\delta_1, \delta_2\}$ then we have the two inequalities

$$\|f(\underline{x} + \underline{h}) - f(\underline{x}) - Df(\underline{x})\underline{h}\| \leq \varepsilon\|\underline{h}\|$$
$$\|g(\underline{z}) - g(\underline{y}) - Dg(\underline{y})(\underline{z} - \underline{y})\| \leq \varepsilon(1 + \|Df(\underline{x})\|)\|\underline{h}\|$$

with $z = f(\underline{x} + \underline{h})$, $y = f(\underline{x})$. So with $z = f(\underline{x} + \underline{h})$, $y = f(\underline{x})$ we see $\|\underline{h}\| < \min\{(1 + \|Df(\underline{x})\|)^{-1}\delta_1, \delta_2\} \Rightarrow$

$$
\begin{aligned}
\|g(z) &- g(y) - Dg(y)Df(\underline{x})\underline{h}\| \\
&= \|g(z) - g(y) - Dg(y)(z - y) - Dg(y)\big(f(\underline{x} + \underline{h}) - f(\underline{x}) - Df(\underline{x})\underline{h}\big)\| \\
&\le \|g(z) - g(y) - Dg(y)(z - y)\| + \|Dg(y)\big(f(\underline{x} + \underline{h}) - f(\underline{x}) - Df(\underline{x})\underline{h}\big)\| \\
&\le \|g(z) - g(y) - Dg(y)(z - y)\| + \|Dg(y)\|\,\|f(\underline{x} + \underline{h}) - f(\underline{x}) - Df(\underline{x})\underline{h}\| \\
&\le \varepsilon\|z - y\| + \varepsilon\|Dg(y)\|\,\|\underline{h}\| \\
&\le \varepsilon\big(1 + \|Df(\underline{x})\| + \|Dg(y)\|\big)\|\underline{h}\| = M\varepsilon\|\underline{h}\|\,,
\end{aligned}
$$

where $M = 1 + \|Df(\underline{x})\| + \|Dg(f(\underline{x}))\|$. That is we have proved that if we take $\delta = \min\{(1 + \|Df(\underline{x})\|)^{-1}\delta_1, \delta_2\}$ then

$$
\|\underline{h}\| < \delta \Rightarrow \|g(f(\underline{x} + h)) - g(f(\underline{x})) - Dg(f(\underline{x}))Df(\underline{x})\underline{h}\| \le M\varepsilon\|\underline{h}\|\,.
$$

Since we can repeat the whole argument with $M^{-1}\varepsilon$ in place of ε we have thus shown that there is a $\delta > 0$ such that

$$
\|\underline{h}\| < \delta \Rightarrow \|g(f(\underline{x} + h)) - g(f(\underline{x})) - Dg(f(\underline{x}))Df(\underline{x})\underline{h}\| \le \varepsilon\|\underline{h}\|\,.
$$

That is, $g \circ f$ is differentiable at $\underline{x}$ and $D(g \circ f)(\underline{x}) = Dg(f(\underline{x}))Df(\underline{x})$.

SECTION 5 EXERCISES

5.1 Use the chain rule and any other theorems from the text that you need to show the following:

(i) If $f : U \to V$ and $g : V \to \mathbb{R}^p$ are C^1, where $U \subset \mathbb{R}^n$, $V \subset \mathbb{R}^m$ are open, then $g \circ f$ is C^1 on U.

(ii) If $f : \mathbb{R} \to \mathbb{R}$ is C^1 and $g : \mathbb{R}^3 \to \mathbb{R}$ is defined by $g(\underline{x}) = f(2x_1 - x_2 + 3x_3)$, prove that g is C^1 and show $D_1 g(\underline{x}) = -2D_2 g(\underline{x}) = (2/3)D_3 g(\underline{x})$ at each point $\underline{x} = (x_1, x_2, x_3) \in \mathbb{R}^3$.

(iii) If $\varphi : \mathbb{R} \to \mathbb{R}$ is C^1 and $g(\underline{x}) = \varphi(\|\underline{x}\|)$ for $\underline{x} \in \mathbb{R}^n$, then g is C^1 on $\mathbb{R}^n \setminus \{\underline{0}\}$ and $D_j g(\underline{x}) = \frac{x_j}{\|\underline{x}\|}\varphi'(\|\underline{x}\|)$ for all $\underline{x} \neq \underline{0}$.

(iv) If $\underline{\gamma} : [0, 1] \to \mathbb{R}^n$ is a C^1 curve and if $f : \mathbb{R}^n \to \mathbb{R}$ is C^1 then $f \circ \underline{\gamma}$ is C^1 on $[0, 1]$ and $(f \circ \underline{\gamma})'(t) = \nabla f(\underline{\gamma}(t)) \cdot \underline{\gamma}'(t) = (D_{\underline{\gamma}'(t)} f)(\underline{\gamma}(t)) \,\forall\, t \in [0, 1]$, where ∇f is the gradient of f (i.e., $(Df)^{\mathrm{T}}$) and $D_{\underline{v}} f$ means the directional derivative of f by $\underline{v}$.

6   HIGHER-ORDER PARTIAL DERIVATIVES

Recall that $f : U \to \mathbb{R}^m$, with $U \subset \mathbb{R}^n$ open, is said to be C^1 on U if all the partial derivatives $D_j f$ exist and are continuous on U. Likewise f is said to be C^2 on U if each of the partial derivatives $D_j f$ is C^1, so in this case the second order partial derivatives $D_i D_j f$ exist and are continuous. An important general fact about such second order partials is that

6.1 $D_i D_j f = D_j D_i f$ on U, provided f is C^2 on U ;

i.e., we claim that the order in which one computes the mixed partials is irrelevant. Of course then if $k \geq 2$ and $f : U \to \mathbb{R}^m$ is C^k (meaning all partial derivatives $D_{i_1} D_{i_2} \cdots D_{i_k} f(\underline{x})$ exist and are continuous for any choice of $i_1, \ldots, i_k \in \{1, \ldots, n\}$), then again the order in which the partial derivatives are taken is irrelevant:

6.2
$$f \in C^k \Rightarrow D_{i_1} D_{i_2} \cdots D_{i_k} f(\underline{x}) = D_{i_{j_1}} D_{i_{j_2}} \cdots D_{i_{j_k}} f(\underline{x})$$

whenever $j_1, \ldots, j_k$ is a re-ordering ("a permutation") of the integers $1, \ldots, k$, which is an easy consequence of 6.1 and induction on k, because for $k \geq 3$ we can write

$$D_{i_1} D_{i_2} \cdots D_{i_k} f(\underline{x}) = D_{i_1} D_{i_2} \cdots D_{i_{k-1}} (D_{i_k} f)(\underline{x}) = D_{i_1} (D_{i_2} \cdots D_{i_{k-1}} D_{i_k} f)(\underline{x}) .$$

A partial derivative $D_{i_1} \cdots D_{i_k} f(\underline{x})$ as in 6.2 above is usually referred to as a *partial derivative of f of order k.*

We will actually prove a result which is slightly more general than 6.1, as follows.

6.3 Theorem. *Suppose $f : U \to \mathbb{R}^m$, where U is an open subset of $\mathbb{R}^n$, suppose $\rho > 0, \underline{a} \in U, B_\rho(\underline{a}) \subset U, i, j \in \{1, \ldots, n\}$, and suppose the second-order partial derivatives $D_i D_j f(\underline{x})$ and $D_j D_i f(\underline{x})$ exist at each point $\underline{x}$ of $B_\rho(\underline{a})$ and are continuous at the point $\underline{x} = \underline{a}$. Then $D_i D_j f(\underline{a}) = D_j D_i f(\underline{a})$.*

Proof: Evidently, there is nothing to prove when $i = j$, so we can assume $i \neq j$. Also, we can assume $m = 1$, because otherwise we can first apply the case $m = 1$ of the theorem to each component f_j of $f = (f_1, \ldots, f_m)^\mathrm{T}$ and this then evidently implies the theorem as stated. So we assume that $m = 1$ (i.e., that f is real-valued) and $i \neq j$. For $h, k \in (-\rho/2, \rho/2)$ we observe that the points $\underline{a} + h\underline{e}_i$, $\underline{a} + k\underline{e}_j$ and $\underline{a} + h\underline{e}_i + k\underline{e}_j$ are all in $B_\rho(\underline{a})$ and so it makes sense to consider the "second difference"

$$f(\underline{a} + h\underline{e}_i + k\underline{e}_j) - f(\underline{a} + h\underline{e}_i) - f(\underline{a} + k\underline{e}_j) + f(\underline{a}) .$$

The expression here is called a second difference because it can be written

$$\varphi_{j,k}(\underline{a} + h\underline{e}_i) - \varphi_{j,k}(\underline{a}) \text{ with } \varphi_{j,k}(\underline{x}) = f(\underline{x} + k\underline{e}_j) - f(\underline{x})$$

(i.e., it can be written as the "first difference of the first difference $\varphi_{j,k}$" hence the name "second difference"), and similarly it can also be written

$$\psi_{i,h}(\underline{a} + k\underline{e}_j) - \psi_{i,h}(\underline{a}) \text{ with } \psi_{i,h}(\underline{x}) = f(\underline{x} + h\underline{e}_i) - f(\underline{x}) ,$$

so in fact,

(1)
$$\varphi_{j,k}(\underline{a} + h\underline{e}_i) - \varphi_{j,k}(\underline{a}) = \psi_{i,h}(\underline{a} + k\underline{e}_j) - \psi_{i,h}(\underline{a}) .$$

Now only the i-th variable is being varied in the difference on the left of (1), so we can use the Mean Value Theorem from 1-variable calculus in order to conclude that

$$\varphi_{j,k}(\underline{a} + h\underline{e}_i) - \varphi_{j,k}(\underline{a}) = h D_i \varphi_{j,k}(\underline{a} + \theta h\underline{e}_i) \text{ for some } \theta \in (0, 1) ,$$

and of course by taking the partial derivative with respect to x_i on each side of the identity $\varphi_{j,k}(\underline{x}) = f(\underline{x} + k\underline{e}_j) - f(\underline{x})$, we see that $D_i \varphi_{j,k}(\underline{x}) = D_i f(\underline{x} + k\underline{e}_j) - D_i f(\underline{x})$. Hence,

$$\varphi_{j,k}(\underline{a} + h\underline{e}_i) - \varphi_{j,k}(\underline{a}) = hD_i f(\underline{a} + \theta h\underline{e}_i + k\underline{e}_j) - hD_i f(\underline{a} + \theta h\underline{e}_i)$$

and on the right here is a difference in which only the j-th variable is being varied, hence we can again use the Mean Value Theorem from 1-variable calculus to see that the expression on the right can be written $hk D_j D_i f(\underline{a} + \theta h\underline{e}_i + \eta k\underline{e}_j)$ for some $\eta \in (0, 1)$. Thus, finally we have shown

(2) $\varphi_{j,k}(\underline{a} + h\underline{e}_i) - \varphi_{j,k}(\underline{a}) = hk D_j D_i f(\underline{a} + \theta h\underline{e}_i + \eta k\underline{e}_j)$ for $|h|, |k| < \rho/2$.

By a similar argument (starting with the right side of (1) instead of the left) we have

(3) $\psi_{i,h}(\underline{a} + k\underline{e}_j) - \psi_{i,h}(\underline{a}) = kh D_i D_j f(\underline{a} + \widetilde{\theta} h\underline{e}_i + \widetilde{\eta} k\underline{e}_j)$ for $|h|, |k| < \rho/2$,

and for suitable $\widetilde{\theta}, \widetilde{\eta} \in (0, 1)$. Thus, by (1), (2), (3) we see that

(4) $D_j D_i f(\underline{a} + \theta h\underline{e}_i + \eta k\underline{e}_j) = D_i D_j f(\underline{a} + \widetilde{\theta} h\underline{e}_i + \widetilde{\eta} k\underline{e}_j)$ for $|h|, |k| < \rho/2$.

Taking limits as $\sqrt{h^2 + k^2} \to 0$ and using the fact that $D_i D_j f(\underline{x})$ and $D_j D_i f(\underline{x})$ are both continuous at $\underline{x} = \underline{a}$, we have $D_i D_j f(\underline{a}) = D_j D_i f(\underline{a})$ as claimed.

SECTION 6 EXERCISES

6.1 Suppose $f, g : \mathbb{R} \to \mathbb{R}$ are C^2 functions, and let $F : \mathbb{R}^2 \to \mathbb{R}$ be defined by $F(x, y) = f(x + y) + g(x - y)$. Prove that F is C^2 on $\mathbb{R}^2$ and satisfies the wave equation on $\mathbb{R}^2$ (i.e., the equation $\frac{\partial^2 F}{\partial x^2} - \frac{\partial^2 F}{\partial y^2} = 0$).

6.2 Prove that the degree n polynomials u, v obtained by taking the real and imaginary parts of $(x + iy)^n$ are harmonic (i.e., $\frac{\partial^2 p}{\partial x^2} + \frac{\partial^2 p}{\partial y^2} \equiv 0$ on $\mathbb{R}^2$ in both cases $p = u, p = v$).

Hint: Thus, $(x + iy)^n = u(x, y) + iv(x, y)$ where u, v are real-valued polynomials in the variables x, y. Start by showing that $\frac{\partial u}{\partial x} = \frac{\partial v}{\partial y}$ and $\frac{\partial u}{\partial y} = -\frac{\partial v}{\partial x}$.

7   THE SECOND DERIVATIVE TEST FOR EXTREMA OF A MULTIVARIABLE FUNCTION

Here we want to discuss conditions on a C^2 function $f : U \to \mathbb{R}$, with U open in $\mathbb{R}^n$, which are sufficient to guarantee f has a local maximum or local minimum at a point $\underline{a} \in U$. First we give the precise definition of local maximum and local minimum.

7.1 Definition: Suppose $f : U \to \mathbb{R}$ with U open and $\underline{a} \in U$. f is said to have a local maximum at $\underline{a}$ if there is $\rho > 0$ with $B_\rho(\underline{a}) \subset U$ and

$$f(\underline{x}) \leq f(\underline{a}) \, \forall \underline{x} \in B_\rho(\underline{a})$$

and the local maximum is said to be *strict* if there is $\rho > 0$ such that $B_\rho(\underline{a}) \subset U$ and

$$f(\underline{x}) < f(\underline{a}) \,\forall\, \underline{x} \in B_\rho(\underline{a}) \setminus \{\underline{a}\} \,.$$

Likewise, f is said to have a local minimum at $\underline{a}$ if there is $\rho > 0$ with $B_\rho(\underline{a}) \subset U$ and $f(\underline{x}) \geq f(\underline{a}) \,\forall\, \underline{x} \in B_\rho(\underline{a})$ and the local minimum is said to be *strict* if $\exists\, \rho > 0$ such that we have the strict inequality $f(\underline{x}) > f(\underline{a}) \,\forall\, \underline{x} \in B_\rho(\underline{a}) \setminus \{\underline{a}\}$.

Of course if $f : U \to \mathbb{R}$ is differentiable with $U \subset \mathbb{R}^n$ open, then $\nabla f(\underline{a}) = \underline{0}$ at any point $\underline{a} \in U$ where f has a local maximum or minimum (because if f has a local maximum or minimum at $\underline{a}$ then $f(\underline{a} + t\underline{e}_j)$ clearly has a local maximum or minimum at $t = 0$ and hence by 1-variable calculus has zero derivative with respect to t at $t = 0$, and the derivative at $t = 0$ is by definition the j'th partial derivative $D_j f(\underline{a})$).

Recall the second derivative test from single variable calculus that if $f : (\alpha, \beta) \to \mathbb{R}$ is C^2 (i.e., the first and second derivatives of f exist and are continuous) then f has a strict local maximum at a point $a \in (\alpha, \beta)$ if $f'(a) = 0$ and $f''(a) < 0$ and a strict local minimum at a point $a \in (\alpha, \beta)$ if $f'(a) = 0$ and $f''(a) > 0$.

Here we want to show that a similar statement holds for functions $f : U \to \mathbb{R}$, with U open in $\mathbb{R}^n$. In order to state the relevant theorem, we first need to make a brief discussion of the notion of *quadratic form*.

A quadratic form on $\mathbb{R}^n$ is an expression of the form

$$7.2 \qquad\qquad Q(\xi) = \sum_{i,j=1}^{n} q_{ij}\xi_i\xi_j, \quad \xi = (\xi_1, \ldots, \xi_n) \in \mathbb{R}^n \,,$$

where (q_{ij}) is a given $n \times n$ real symmetric matrix. The quadratic form is said to be *positive definite* if $Q(\xi) > 0 \,\forall\, \xi \in \mathbb{R}^n \setminus \{\underline{0}\}$ and *negative definite* if $Q(\xi) < 0 \,\forall\, \xi \in \mathbb{R}^n \setminus \{\underline{0}\}$.

Of course since a quadratic form is just a homogeneous degree 2 polynomial in the variables $\xi_1, \ldots, \xi_n$, it is evidently continuous and hence by Thm. 2.6 of the present chapter attains its maximum and minimum value on any closed bounded set in $\mathbb{R}^n$. In particular, the unit sphere S^{n-1} in $\mathbb{R}^n$,

$$7.3 \qquad\qquad S^{n-1} = \{\underline{x} \in \mathbb{R}^n : \|\underline{x}\| = 1\} \,,$$

is a closed bounded set, hence the restriction of the quadratic form to S^{n-1}, $Q|S^{n-1}$, therefore does attain its maximum and minimum value. Let

$$7.4 \qquad\qquad m = \min Q|S^{n-1}, \quad M = \max Q|S^{n-1}$$

and observe that for $\xi \in \mathbb{R}^n \setminus \{\underline{0}\}$ we can write $\xi = \|\xi\|\widehat{\xi}$, where $\widehat{\xi} = \|\xi\|^{-1}\xi \in S^{n-1}$, and so $Q(\xi) = \|\xi\|^2 Q(\widehat{\xi})$, whence

$$7.5 \qquad\qquad m\|\xi\|^2 \leq Q(\xi) \leq M\|\xi\|^2, \quad \forall\, \xi \in \mathbb{R}^n \,,$$

with m, M as in 7.4.

Notice, in particular, if Q is positive definite then the minimum m of $Q|S^{n-1}$ is positive, and so 7.5 guarantees

7.6 $\qquad Q$ positive definite $\Rightarrow \exists m > 0$ such that $Q(\xi) \geq m\|\xi\|^2 \; \forall \xi \in \mathbb{R}^n$,

and similarly,

7.7 $\qquad Q$ negative definite $\Rightarrow \exists m > 0$ such that $Q(\xi) \leq -m\|\xi\|^2 \; \forall \xi \in \mathbb{R}^n$.

We are now ready to state the second derivative test for C^2 functions $f : U \to \mathbb{R}$, where U is open in $\mathbb{R}^n$. In this theorem, $\mathrm{Hess}_{f,\underline{a}}(\xi)$ will denote the Hessian quadratic form of f at $\underline{a}$, defined by

7.8 $$\mathrm{Hess}_{f,\underline{a}}(\xi) = \sum_{i,j=1}^n D_i D_j f(\underline{a})\xi_i\xi_j, \quad \xi = (\xi_1,\ldots,\xi_n)^{\mathrm{T}} \in \mathbb{R}^n .$$

7.9 Theorem (Second derivative test for functions of n variables.) *Suppose U is open in $\mathbb{R}^n$, $f : U \to \mathbb{R}$ is C^2, and $\underline{a} \in U$. Then*
$$f(\underline{x}) = f(\underline{a}) + (\underline{x} - \underline{a}) \cdot \nabla f(\underline{a}) + \tfrac{1}{2}\mathrm{Hess}_{f,\underline{a}}(\underline{x} - \underline{a}) + E(\underline{x}) ,$$
where $\lim_{\underline{x}\to\underline{a}} \|\underline{x} - a\|^{-2}E(\underline{x}) = 0$.

In particular, f has a strict local maximum at $\underline{a}$ if $\nabla f(\underline{a}) = \underline{0}$ and $\mathrm{Hess}_{f,\underline{a}}(\xi)$ is negative definite, and a strict local minimum at $\underline{a}$ if $\nabla f(\underline{a}) = \underline{0}$ and $\mathrm{Hess}_{f,\underline{a}}(\xi)$ is positive definite.

7.10 Terminology: If $f : U \to \mathbb{R}$ is C^1 and $U \subset \mathbb{R}^n$ is open, points $\underline{a} \in U$ with $\nabla f(\underline{a}) = \underline{0}$ are usually referred to as critical points of f. Using this terminology, Thm. 7.9 says that if f is C^2 then f has a strict local minimum at any critical point $\underline{a} \in U$ where $\mathrm{Hess}_{f,\underline{a}}(\xi)$ is positive definite and strict local maximum at any critical point $\underline{a} \in U$ where $\mathrm{Hess}_{f,\underline{a}}(\xi)$ is negative definite.

Proof of 7.9: We begin with an identity from single variable calculus: If $h : [0, 1] \to \mathbb{R}$ is C^2, then we have the identity

(1) $\qquad \int_0^1 (1 - t)h''(t)\, dt = h(1) - h(0) - h'(0)$,

which is proved using integration by parts. (See Exercise 7.3 below.) We apply (1) in the special case when $h(t) = f(\underline{a} + t(\underline{x} - \underline{a}))$, with $\underline{x} \in B_\rho(\underline{a})$. By the Chain Rule, $h(t)$ is then indeed C^2 on $[0, 1]$ and $h'(t) = (\underline{x} - \underline{a}) \cdot \nabla f(\underline{a} + t(\underline{x} - \underline{a}))$, $h''(t) = \sum_{i,j=1}^n (x_i - a_i)(x_j - a_j)D_i D_j f(\underline{a} + t(\underline{x} - \underline{a}))$, so (1) gives

$f(\underline{x}) = f(\underline{a}) + (\underline{x} - \underline{a}) \cdot \nabla f(\underline{a}) + \sum_{i,j=1}^n (x_i - a_i)(x_j - a_j)\int_0^1 (1 - t)D_i D_j f(\underline{a} + t(\underline{x} - \underline{a}))\, dt$.

Since $\int_0^1 (1 - t)D_i D_j f(\underline{a} + t(\underline{x} - \underline{a}))\, dt = \int_0^1 (1 - t)(D_i D_j f(\underline{a} + t(\underline{x} - \underline{a})) - D_i D_j f(\underline{a}))\, dt + \int_0^1 (1 - t)\, dt\, D_i D_j f(\underline{a})$ and $\int_0^1 (1 - t)\, dt = \tfrac{1}{2}$, this can be rewritten

(2) $\qquad f(\underline{x}) = f(\underline{a}) + (\underline{x} - \underline{a}) \cdot \nabla f(\underline{a}) + \tfrac{1}{2}\mathrm{Hess}_{f,\underline{a}}(\underline{x} - \underline{a}) + E(\underline{x}) ,$

where $E(\underline{x}) = \sum_{i,j=1}^{n} q_{ij}(\underline{x})(x_i - a_i)(x_j - a_j)$ with

$$q_{ij}(\underline{x}) = \int_0^1 (1-t)\big(D_i D_j f(\underline{a} + t(\underline{x} - \underline{a}) - D_i D_j f(\underline{a})\big)\, dt \ .$$

Now observe that

(3)
$$\begin{aligned}
|q_{ij}(\underline{x})| &= \big|\int_0^1 (1-t)\big(D_i D_j f(\underline{a} + t(\underline{x} - \underline{a}) - D_i D_j f(\underline{a})\big)\, dt\big| \\
&\leq \int_0^1 (1-t)|D_i D_j f(\underline{a} + t(\underline{x} - \underline{a}) - D_i D_j f(\underline{a})|\, dt \ ,
\end{aligned}$$

and let $\varepsilon > 0$ and $i, j \in \{1, \ldots, n\}$. Since $D_i D_j f(\underline{x})$ is continuous at $\underline{x} = \underline{a}$ we then have a $\delta_{ij} > 0$ such that $\underline{x} \in B_{\delta_{ij}}(\underline{a}) \Rightarrow |D_i D_j f(\underline{x}) - D_i D_j f(\underline{a})| < \varepsilon/n^2$. Let $\delta = \min\{\delta_{ij} : i, j \in \{1, \ldots, n\}\}$ and observe that $\underline{x} \in B_\delta(\underline{a}) \Rightarrow \underline{a} + t(\underline{x} - \underline{a}) \in B_\delta(\underline{a}) \, \forall\, t \in [0, 1]$, so then we have

$$\underline{x} \in B_\delta(\underline{a}) \Rightarrow |D_i D_j f(\underline{a} + t(\underline{x} - \underline{a}) - D_i D_j f(\underline{a})| < \varepsilon/n^2 \text{ for each } i, j = 1, \ldots, n$$

and each $t \in [0, 1]$. Hence, using (3),

$$\underline{x} \in B_\delta(\underline{a}) \Rightarrow |q_{ij}(\underline{x})| \leq \int_0^1 (1-t)\, dt\, \varepsilon/n^2 = \tfrac{1}{2}\varepsilon/n^2 < \varepsilon/n^2 \ ,$$

and so

(4) $$\underline{x} \in B_\delta(\underline{a}) \Rightarrow |E(\underline{x})| \leq \sum_{i,j=1}^{n} |x_i - a_i||x_j - a_j||q_{ij}(\underline{x})| \leq \|\underline{x} - \underline{a}\|^2 n^2 \varepsilon/n^2 = \varepsilon\|\underline{x} - \underline{a}\|^2 \ .$$

Notice that this is the ε, δ definition of $\lim_{\underline{x} \to \underline{a}} \|\underline{x} - \underline{a}\|^{-2} E(\underline{x}) = 0$, so the first part of the lemma is proved.

Now suppose that $\mathrm{Hess}_{f,\underline{a}}(\xi)$ is positive definite. Then by 7.6 there is $m > 0$ such that the term $\frac{1}{2} \mathrm{Hess}_{f,\underline{a}}(\underline{x} - \underline{a})$ in (2) satisfies

(5)
$$\tfrac{1}{2} \mathrm{Hess}_{f,\underline{a}}(\underline{x} - \underline{a}) \geq m\|\underline{x} - \underline{a}\|^2 \text{ for all } \underline{x} \in \mathbb{R}^n \ ,$$

and, on the other hand, by using (4) with $\varepsilon = m/2$ we have $\delta > 0$ such that

(6) $$\underline{x} \in B_\delta(\underline{a}) \Rightarrow |E(\underline{x})| \leq m\|\underline{x} - \underline{a}\|^2/2 \Rightarrow E(\underline{x}) \geq -m\|\underline{x} - \underline{a}\|^2/2 \ ,$$

and so using (5) and (6) in (2) we see that if $\nabla f(\underline{a}) = \underline{0}$

$$\underline{x} \in B_\delta(\underline{a}) \Rightarrow f(\underline{x}) \geq f(\underline{a}) + m\|\underline{x} - \underline{a}\|^2/2 \ ,$$

so f has a strict local minimum at $\underline{a}$.

Applying the same argument with $-f$ in place of f we deduce that f has a strict local maximum at $\underline{a}$ if $\nabla f(\underline{a}) = \underline{0}$ and if $\mathrm{Hess}_{f,\underline{a}}(\xi)$ is negative definite.

SECTION 7 EXERCISES

7.1 (a) We proved above that if $U \subset \mathbb{R}^n$ is open, if $f : U \to \mathbb{R}$ is C^2, and if $\underline{x}_0 \in U$ is a critical point (i.e., $\nabla f(\underline{x}_0) = \underline{0}$), then f has a local minimum/maximum at $\underline{x}_0$ if the Hessian quadratic form $Q(\xi) \equiv \sum_{i,j=1}^{n} D_i D_j f(\underline{x}_0)\xi_i \xi_j$ is positive definite/negative definite, respectively. Prove that f has neither a local maximum nor a local minimum at a critical point $\underline{x}_0$ of f if $Q(\xi)$ is *indefinite* (i.e., if there are unit vectors ξ, η such that $Q(\xi) > 0$ and $Q(\eta) < 0$.

(b) Find all local max/local min for the function $f : \mathbb{R}^2 \to \mathbb{R}$ given by $f(x, y) = \frac{1}{3}(x^3 + y^3) - x^2 - 2y^2 - 3x + 3y$, and justify your answer.

7.2 Suppose that $f(x, y) = \frac{1}{3}(x^3 + y^3) - x^2 - 2y^2 - 3x + 3y$. Find the critical points (i.e., the points where $\nabla f = \underline{0}$) of f, and discuss whether f has a local maximum/minimum at these points. (Justify any claims that you make by proof or by referring to the relevant theorem above.)

7.3 Prove the identity $\int_0^1 (1-t)h''(t)\,dt = h(1) - h(0) - h'(0)$ used in the proof of Thm. 7.9, assuming that h is a C^2 function on $[0, 1]$.

Hint: Write $\int_0^1 (1-t)h''(t)\,dt = \int_0^1 (1-t)\frac{d}{dt}h'(t)\,dt$, and use integration by parts.

8   CURVES IN $\mathbb{R}^n$

By a C^0 curve in $\mathbb{R}^n$ we mean a continuous map $\gamma : [a, b] \to \mathbb{R}^n$, where $a, b \in \mathbb{R}$ with $a < b$. Of course one might think geometrically of the curve as the *image* $\gamma([a, b])$, but formally the curve is in fact the map $\gamma : [a, b] \to \mathbb{R}^n$.

By a C^1 curve in $\mathbb{R}^n$ we shall mean a C^1 map $\gamma : [a, b] \to \mathbb{R}^n$. We should clarify this, because to date we have only considered functions which are C^1 on open sets, whereas here we take the domain to be the closed interval $[a, b]$. In fact, by saying that $\gamma : [a, b] \to \mathbb{R}^n$ is C^1 we mean that: (i) the derivative $\gamma'(t) = \lim_{h \to 0} h^{-1}(\gamma(t+h) - \gamma(t))$ exists for each $t \in (a, b)$; (ii) the *one-sided* derivatives $\gamma'(a) = \lim_{h \downarrow 0} h^{-1}(\gamma(a+h) - \gamma(a))$, $\gamma'(b) = \lim_{h \uparrow 0} h^{-1}(\gamma(a+h) - \gamma(a))$ exist; and (iii) $\gamma'(t)$, so defined, is a continuous function of t on the closed interval $[a, b]$.

$\gamma'(t)$ is sometimes referred to as the *velocity vector* of γ, because if $t, t+h \in [a, b]$, the quantity $\gamma(t+h) - \gamma(t)$ is the change of position of γ as t changes from t to $t+h$, so, as $h \to 0, h^{-1}(\gamma(t+h) - \gamma(t))$ can be thought of as the (vector) rate of change of the point $\gamma(t)$. If $\gamma'(t) \neq \underline{0}$ we define

8.1 $$\tau(t) = \|\gamma'(t)\|^{-1}\gamma'(t),$$

and we refer to $\tau(t)$ as the unit tangent vector of the curve γ. From the schematic diagram (depicting the case $n = 3$) it is intuitively evident that indeed $\gamma'(t)$ is tangent to the curve $\gamma(t)$ if it is nonzero, and hence that the terminology "unit tangent vector" is appropriate.

We can also discuss the *length* $\ell(\gamma)$ of a C^0 curve γ as follows.

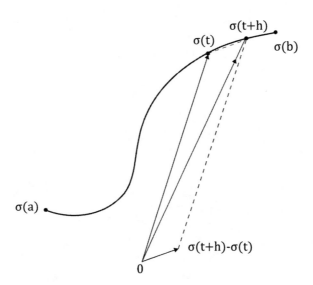

Figure 2.2: Schematic picture of definition of velocity vector $n = 3$.

First we introduce the notion of *partition* of the interval $[a, b]$, meaning a finite collection $\mathcal{P} = \{t_0, \ldots, t_N\}$ of points of $[a, b]$ with $a = t_0 < t_1 < \cdots < t_N = b$. For such a partition $\mathcal{P}$ we define

$$\ell(\gamma, \mathcal{P}) = \sum_{j=1}^{N} \|\gamma(t_j) - \gamma(t_{j-1})\| .$$

Geometrically, $\ell(\gamma, \mathcal{P})$ is the length of the "polygonal approximation" to γ obtained by taking the union of the straight line segments $\overline{\gamma(t_{j-1})\,\gamma(t_j)}$ joining the points $\gamma(t_{j-1})$ and $\gamma(t_j)$, as in the following schematic diagram, which depicts the case $n = 3$.

We say the curve has *finite length* if the set

8.2 $$\{\ell(\gamma, \mathcal{P}) : \mathcal{P} \text{ is a partition of } [a, b]\}$$

is bounded above. In case γ has finite length we can define the length $\ell(\gamma)$ to be the least upper bound of the set in 8.2:

8.3 $$\ell(\gamma) = \sup\{\ell(\gamma, \mathcal{P}) : \mathcal{P} \text{ is a partition of } [a, b]\} .$$

It is a straightforward exercise (Exercise 8.1 below) to check the additivity property that if $c \in (a, b)$ then

8.4 $$\ell(\gamma) = \ell(\gamma|[a, c]) + \ell(\gamma|[c, b]) .$$

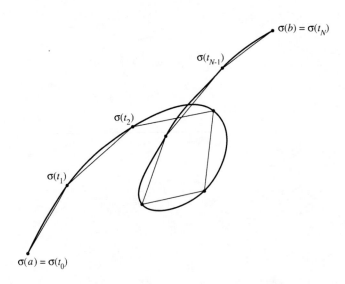

Figure 2.3: Schematic picture of polygonal approximation of γ, $n = 3$.

The explicit formula for length of a C^1 curve given in the following theorem is important.

8.5 Theorem. *If $\gamma : [a, b] \to \mathbb{R}^n$ is C^1, then γ has finite length and in fact*

$$\ell(\gamma) = \int_a^b \|\gamma'(t)\| \, dt .$$

Proof: First observe by the fundamental theorem of calculus

$$\gamma(t_j) - \gamma(t_{j-1}) = \int_{t_{j-1}}^{t_j} \gamma'(t) \, dt .$$

Using the fact that $\| \int_{t_{j-1}}^{t_j} \gamma'(t) \, dt \| \le \int_{t_{j-1}}^{t_j} \|\gamma'(t)\| \, dt$ (see Exercise 8.3 below), we then have

$$\|\gamma(t_j) - \gamma(t_{j-1})\| \le \int_{t_{j-1}}^{t_j} \|\gamma'(t)\| \, dt$$

for each $j = 1, \ldots, N$, and hence by summing over j we have

$$\sum_{j=1}^N \|\gamma(t_j) - \gamma(t_{j-1})\| \le \int_a^b \|\gamma'(t)\| \, dt ,$$

so the collection of all sums

$$\{ \sum_{j=1}^N \|\gamma(t_j) - \gamma(t_{j-1})\| : t_0 = a < t_1 < \cdots < t_N = b \text{ is a partition of } [a, b] \}$$

has the integral $\int_a^b \|\gamma'(t)\| \, dt$ as an upper bound so indeed γ has finite length as claimed and

(1)
$$\ell(\gamma) \leq \int_a^b \|\gamma'(t)\| \, dt \, .$$

For $t \in [a, b]$ we define
$$S(t) = \ell(\gamma|[a, t])$$

so that $S(a) = 0$ and $S(b) = \ell(\gamma)$. Take $\tau \in [a, b)$ and $h > 0$ such that $\tau + h \in [a, b]$, and observe that, by 8.4 with $a, \tau, \tau + h$ in place of a, c, b, respectively, we have
$$S(\tau + h) = S(\tau) + \ell(\gamma|[\tau, \tau + h])$$

so

(2)
$$\ell(\gamma|[\tau, \tau + h]) = S(\tau + h) - S(\tau) \, ,$$

but, on the other hand, by (1) and the definition of length we have
$$\|\gamma(\tau + h) - \gamma(\tau)\| \leq \ell(\gamma|[\tau, \tau + h]) \leq \int_\tau^{\tau+h} \|\gamma'(t)\| \, dt \, ,$$

i.e., by (2)
$$\|\gamma(\tau+h) - \gamma(\tau)\| \leq S(\tau+h) - S(\tau) \leq \int_\tau^{\tau+h} \|\gamma'(t)\| \, dt = \int_a^{\tau+h} \|\gamma'(t)\| \, dt - \int_a^\tau \|\gamma'(t)\| \, dt \, ,$$

so
$$h^{-1} \|\gamma(\tau + h) - \gamma(\tau)\| \leq h^{-1}(S(\tau + h) - S(\tau)) \leq h^{-1}\left(\int_a^{\tau+h} \|\gamma'(t)\| \, dt - \int_a^\tau \|\gamma'(t)\| \, dt \right) .$$

Notice the right side has limit as $h \downarrow 0$ equal to $\frac{d}{d\tau} \int_a^\tau \|\gamma'(t)\| \, dt$, which by the fundamental theorem of calculus is $\|\gamma'(\tau)\|$. So both the left and right side has limit as $h \downarrow 0$ equal to $\|\gamma'(\tau)\|$, and hence by the Sandwich Theorem (Exercise 2.1 of the present chapter) we deduce that for $\tau \in [a, b)$
$$\lim_{h \downarrow 0} h^{-1}(S(\tau + h) - S(\tau)) = \|\gamma'(\tau)\| \, .$$

By a similar argument with $h < 0$ we also prove for $\tau \in (a, b]$
$$\lim_{h \uparrow 0} h^{-1}(S(\tau + h) - S(\tau)) = \|\gamma'(\tau)\| \, .$$

Thus, we have shown $s'(t)$ exists and equals $\|\gamma'(t)\|$ for every $t \in [a, b]$,[1] so by the fundamental theorem of calculus and the fact that $S(a) = 0$ we have
$$S(b) = \int_a^b \|\gamma'(t)\| \, dt \, .$$

[1]Of course we proved only the existence of the relevant 1-sided derivatives at the end-points a, b.

Since, by definition $S(b) = \ell(\gamma)$, this completes the proof.

8.6 Remark: If $\gamma : [a, b] \to \mathbf{R}^n$ is C^1, then, as shown in the above proof, the function $S(t) = \ell(\gamma|[a, t])$ is a C^1 function on $[a, b]$, and $S'(t) = \|\gamma'(t)\|$ for each $t \in [a, b]$. Thus, if $\gamma'(t) \neq 0$ then S has positive derivative, hence is a strictly increasing function mapping $[a, b]$ onto $[0, \ell(\gamma)]$, and by 1-variable calculus the inverse function $T : [0, \ell(\gamma)] \to [a, b]$ is C^1 and the derivative $T'(s) = (S'(t))^{-1}$, assuming the variables s, t are related by $s = S(t)$ (or, equivalently, $t = T(s)$). Notice that then if we define $\tilde{\gamma} = \gamma \circ T$, so $\gamma(t) = \tilde{\gamma}(s)$, assuming again that s, t are related by $s = S(t)$ or, equivalently, $t = T(s)$. The map $\tilde{\gamma}$ is called the arc-length parametrization of γ (because $s = S(t) = \ell(\gamma|[a, t])$). Notice that by the 1-variable calculus chain rule we have $\tilde{\gamma}'(s) = \|\gamma'(t)\|^{-1}\gamma'(t)$ and, in particular, $\|\tilde{\gamma}'(s)\| \equiv 1$. The reader should keep in mind that all of this is dependent on the assumption that γ is C^1 with $\gamma'(t) \neq 0$.

SECTION 8 EXERCISES

8.1 Let $\gamma : [a, b] \to \mathbf{R}^n$ be an arbitrary continuous curve of finite length. Prove directly from the definition of length (as the least upper bound of lengths of polygonal approximations, as in 8.3) that if $c \in (a, b)$ then

$$\ell(\gamma|[a, c]) + \ell(\gamma|[c, b]) = \ell(\gamma).$$

Hint: Start by showing that if $\mathcal{P}, \mathcal{Q}$ are partitions for $[a, c]$ and $[c, b]$, respectively, then $\ell(\gamma|[a, c], \mathcal{P}) + \ell(\gamma|[c, b], \mathcal{Q}) \leq \ell(\gamma)$.

8.2 Let $\gamma : [0, 1] \to \mathbf{R}^2$ be the C^0 curve defined by $\gamma(t) = (t, t \sin \frac{1}{t})$ if $t \in (0, 1]$ and $\gamma(0) = (0, 0)$. Prove that γ has infinite length.

Hint: One approach is to directly check that for each real $K > 0$ there is a partition $\mathcal{P} : t_0 = 0 < t_1 < t_2 < \cdots < t_N = 1$ such that $\sum_{j=1}^{N} \|\gamma(t_j) - \gamma(t_{j-1})\| > K$.

8.3 Let $f = (f_1, \ldots, f_n) : [a, b] \to \mathbf{R}^n$ be continuous and as usual take $\int_a^b f(t)\,dt$ to mean $\left(\int_a^b f_1(t)\,dt, \ldots, \int_a^b f_n(t)\,dt\right)(\in \mathbf{R}^n)$. Prove $\|\int_a^b f(t)\,dt\| \leq \int_a^b \|f(t)\|\,dt$. Hint: First observe $\underline{v} \cdot \int_a^b f(t)\,dt = \int_a^b \underline{v} \cdot f(t)\,dt$ for any fixed vector $\underline{v} \in \mathbf{R}^n$.

8.4 Consider the "helix" in $\mathbf{R}^3$ given by the curve $\gamma(t) = (a \cos \omega t, a \sin \omega t, b\omega t)$ for $t \in \mathbf{R}$, where a, b, ω are given positive constants.

(a) Show that the speed (i.e., $\|\gamma'(t)\|$) is constant for $t \in \mathbf{R}$.

(b) Show that the velocity vector $\gamma'(t)$ makes a constant nonzero angle with the z-axis.

(c) With $t_1 = 0$ and $t_2 = 2\pi/\omega$, show that there is no $\tau \in (t_1, t_2)$ such that $\gamma(t_1) - \gamma(t_2) = (t_1 - t_2)\gamma'(\tau)$.

(Thus the Mean Value Theorem fails for vector-valued functions.)

8.5 For the Helix $\gamma(t) = (a \cos \omega t, a \sin \omega t, b\omega t)$, where $t \in [0, 2\pi/\omega]$ and $a, b, \omega > 0$, prove that the arc-length variable $s = S(t) (= \ell(\gamma|[0, t]))$ (as in Rem. 8.6 above) is a constant multiple of t. Also, find the length, unit tangent vector, and curvature vector $\widetilde{\gamma}''(s)$ for γ.

8.6 Suppose $\gamma : [a, b] \to \mathbb{R}^n$ is a C^2 curve with nonzero velocity vector γ'. If $\widetilde{\gamma}(s) = \gamma(t)$, where s is the arc-length parameter (i.e., $s = S(t) = \ell(\gamma|[a, t])$ as in Rem. 8.6 above), prove that the curvature vector $\kappa = \frac{d}{ds}\widetilde{\gamma}'(s)$ at s is given in terms of the t-variable by the formula

$$\kappa(s) = \|\gamma'(t)\|^{-2}\Big(I - \|\gamma'(t)\|^{-2}\gamma'(t)\gamma'(t)^{\mathrm{T}}\Big)\gamma''(t) .$$

Here I is the $n \times n$ identity matrix and we assume $\gamma(t)$ is written as a column; $\gamma(t)\gamma(t)^{\mathrm{T}}$ is matrix multiplication of the $n \times 1$ matrix $\gamma(t)$ by the $1 \times n$ matrix $\gamma(t)^{\mathrm{T}}$.

9   SUBMANIFOLDS OF R^n AND TANGENTIAL GRADIENTS

In this section it will be convenient to write vectors in $\mathbb{R}^n$ as rows, $\underline{x} = (x_1, \ldots, x_n)$, and the reader should keep in mind that then the gradient $\nabla f(\underline{x})$ of a C^1 function $f : U \to \mathbb{R}$ will also be a row vector $\nabla f(\underline{x}) = (D_1 f(\underline{x}), \ldots, D_n f(\underline{x}))$.

We first introduce the notion of k-dimensional graph in $\mathbb{R}^n$, with $k \in \{1, \ldots, n-1\}$.

9.1 Definition: If $g = (g_1, \ldots, g_{n-k}) : U \to \mathbb{R}^{n-k}$, where U is open in $\mathbb{R}^k$, *the graph map* corresponding to g is the map $G : U \to \mathbb{R}^n$ defined by

$$G(\underline{x}) = (\underline{x}, g(\underline{x})), \quad \underline{x} \in U ,$$

and the *graph of g* is defined to be $G(U)$, i.e., $\{G(\underline{x}) : \underline{x} \in U\}$. Notice that G is C^1 if g is C^1 and in this case we can view the graph of g (i.e., $G(U)$) as a k-dimensional C^1 surface sitting in $\mathbb{R}^n$, as represented in the following schematic diagram, which is actually representative of the special case $k = 2, n = 3$.

9.2 Definition: A subset $M \subset \mathbb{R}^n$ is a k-dimensional C^1 submanifold of $\mathbb{R}^n$ if for each $\underline{a} \in M$ there is a $\delta > 0$ such that $M \cap B_\delta(\underline{a})$ is $P_{j_1,\ldots,j_n}(G(U))$ where $G(\underline{x})$ (as in 9.1) is the graph map $(\underline{x}, g(\underline{x}))$ of a C^1 function $g : U \to \mathbb{R}^{n-k}$ with U open in $\mathbb{R}^k$, and $P_{j_1,\ldots,j_n}$ is the permutation map $P_{j_1,\ldots,j_n}(x_1, \ldots, x_n) = (x_{j_1}, \ldots, x_{j_n})$, with $j_1, \ldots, j_n$ a permutation (i.e., a re-ordering) of $1, \ldots, n$.

This definition sounds at first a little abstract, but it is really just saying that if $\underline{a} \in M$ then, in a suitably small ball $B_\delta(\underline{a})$, we can re-label the coordinates (using x_{j_i} as the new i-th coordinate for $i = 1, \ldots, n$) in such a way that $M \cap B_\delta(\underline{a})$ becomes the graph of a C^1 function as in 9.1.

Figure 2.5 gives a schematic picture of what is going on in the case $k = 1, n = 2$.

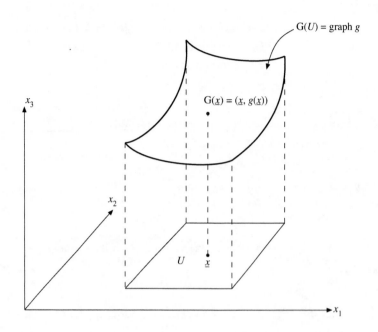

Figure 2.4: Schematic picture of a C^1 graph in case $k = 2, n = 3$.

Next we define the notion of tangent space $T_{\underline{a}} M$ of the C^1 manifold M at $\underline{a} \in M$, as follows.

9.3 Definition: If M is a k-dimensional C^1 submanifold of $\mathbb{R}^n$ and $\underline{a} \in M$ then the tangent space of M at $\underline{a}$, $T_{\underline{a}} M$, is defined by

$$T_{\underline{a}} M = \{\gamma'(0) : \gamma \text{ is a } C^1 \text{ map } [0, \alpha) \to \mathbb{R}^n \text{ with } \alpha > 0, \ \gamma([0, \alpha)) \subset M, \gamma(0) = \underline{a}\} \ .$$

Thus, in fact $T_{\underline{a}} M$ is the set of all initial velocity vectors of C^1 curves $\gamma(t)$ which are contained in M and which start from the given point $\underline{a}$ at $t = 0$. We claim that $T_{\underline{a}} M$ is a k-dimensional subspace of $\mathbb{R}^n$.

9.4 Lemma. *If M is a k-dimensional C^1 submanifold of $\mathbb{R}^n$ and if $\underline{a} \in M$, then the tangent space $T_{\underline{a}} M$ is a k-dimensional subspace of $\mathbb{R}^n$.*

9.5 Remark A schematic picture representing the situation described in the above lemma, in the case $k = 2$ and $n = 3$, is given in Figure 2.6.

Proof of 9.4: Since it is merely a matter of relabeling the coordinates, we can suppose without loss of generality that the permutation $j_1, \ldots, j_n$ needed in the Def. 9.2 is in fact the identity permutation (i.e., $j_i = i$ for each $i = 1, \ldots, n$), and hence there is $\delta > 0$ such that

$$M \cap B_\delta(\underline{a}) = \text{graph } g \ ,$$

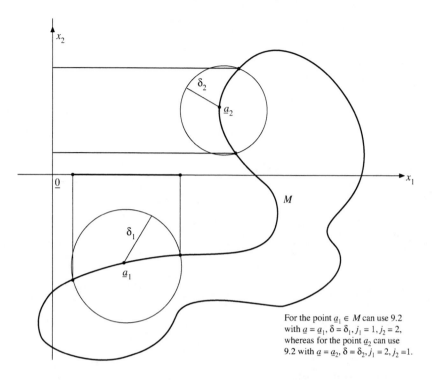

For the point $\underline{a}_1 \in M$ can use 9.2
with $\underline{a} = \underline{a}_1$, $\delta = \delta_1$, $j_1 = 1$, $j_2 = 2$,
whereas for the point $\underline{a}_2$ can use
9.2 with $\underline{a} = \underline{a}_2$, $\delta = \delta_2$, $j_1 = 2$, $j_2 = 1$.

Figure 2.5: Schematic picture of a C^1 manifold M in case $k = 1, n = 2$.

where $g : U \to \mathbb{R}^{n-k}$ is C^1 and U is open in $\mathbb{R}^k$. If $\underline{v} \in T_{\underline{a}} M$ then by definition of $T_{\underline{a}} M$ there is a C^1 curve $\gamma : [0, \alpha) \to \mathbb{R}^n$ ($\alpha > 0$) with $\gamma([0, \alpha)) \subset M$, $\gamma(0) = \underline{a}$ and $\gamma'(0) = \underline{v}$. Since γ is continuous we have $\rho > 0$ such that $0 \le t < \rho \Rightarrow \gamma(t) \in M \cap B_\delta(\underline{a})$, and so $0 \le t < \rho \Rightarrow$

$$(1) \qquad\qquad \gamma(t) = G(\sigma(t)) ,$$

where G is the graph map as in 9.1 and $\sigma(t) = (\gamma_1(t), \ldots, \gamma_k(t))$ (i.e., $\sigma(t)$ is the vector in $\mathbb{R}^k$ obtained by taking first k components of $\gamma(t)$), so that $\sigma(t) : [0, \rho) \to \mathbb{R}^k$ is C^1 with $\sigma(0) = \underline{a}_0$, where $\underline{a}_0 = (a_1, \ldots, a_k)$ (i.e., $\underline{a}_0$ is the vector in $\mathbb{R}^k$ obtained by taking the first k components of $\underline{a}$). By applying the chain rule to the composite on the right of (1) we then have

$$(2) \qquad\qquad \underline{v} = \gamma'(0) = \sum_{j=1}^{n} y_j D_j G(\underline{a}_0) ,$$

where $(y_1, \ldots, y_n) = \sigma'(0) \in \mathbb{R}^k$, thus showing, since $\underline{v} \in T_{\underline{a}} M$ was arbitrary, that

$$(3) \qquad\qquad T_{\underline{a}} M \subset \text{span}\{D_1 G(\underline{a}_0), \ldots, D_k G(\underline{a}_0)\} .$$

We can also prove the reverse inclusion

$$(4) \qquad\qquad \text{span}\{D_1 G(\underline{a}_0), \ldots, D_k G(\underline{a}_0)\} \subset T_{\underline{a}} M ,$$

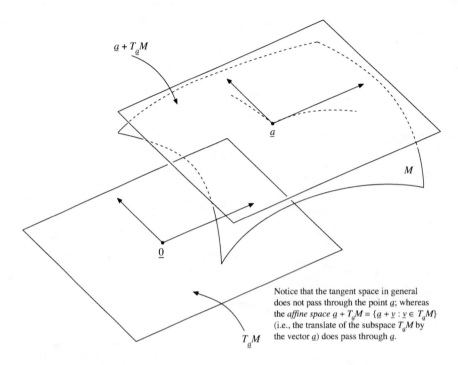

Notice that the tangent space in general does not pass through the point $\underline{a}$; whereas the *affine space* $\underline{a} + T_{\underline{a}}M = \{\underline{a} + \underline{v} : \underline{v} \in T_{\underline{a}}M\}$ (i.e., the translate of the subspace $T_{\underline{a}}M$ by the vector $\underline{a}$) does pass through $\underline{a}$.

Figure 2.6: The tangent space $T_{\underline{a}}M$ of M.

Notice that the tangent space in general does not pass through the point $\underline{a}$; whereas the *affine space* $\underline{a} + T_{\underline{a}}M = \{\underline{a} + \underline{v} : \underline{v} \in T_{\underline{a}}M\}$ (i.e., the translate of the subspace $T_{\underline{a}}M$ by the vector $\underline{a}$) does pass through $\underline{a}$.

because if $\underline{y} \in \mathbb{R}^k$ we can take $\rho > 0$ such that $B_\rho(\underline{a}_0) \subset U$ (because U is open), and hence for any $\alpha > 0$ with $\alpha \|\underline{y}\| < \rho$ we have $\gamma(t) = G(\underline{a}_0 + t\underline{y})$ is C^1 for $|t| < \alpha$, $\gamma((-\alpha, \alpha)) \subset M$, $\gamma(0) = \underline{a}$ and again (2) holds by the chain rule. Since $y_1, \ldots, y_k$ were arbitrary, this does prove the reverse inclusion (4). Thus, in fact (by (3),(4)) we have proved

(5) $$T_{\underline{a}}M = \operatorname{span}\{D_1 G(\underline{a}_0), \ldots, D_k G(\underline{a}_0)\} \ .$$

Also, $D_1 G(\underline{a}_0), \ldots, D_k G(\underline{a}_0)$ are l.i.; indeed,

$$G(\underline{x}) = (\underline{x}, g(\underline{x})) \Rightarrow D_j G(\underline{x}) = (\underline{e}_j, D_j g(\underline{x})), \ j = 1, \ldots, k \ ,$$

which are clearly l.i. vectors. This completes the proof that $T_{\underline{a}}M$ is a k-dimensional subspace of $\mathbb{R}^n$.

9.5 Remark: Part of the above proof (in fact the proof of inclusion (4)) shows that for any $\underline{a} \in M$ and for any $\underline{y} \in \mathbb{R}^k$ we can find $\gamma : (-\alpha, \alpha) \to \mathbb{R}^n$ (i.e., γ is C^1 on the whole interval $(-\alpha, \alpha)$ rather than just on $[0, \alpha)$) with $\alpha > 0$, $\gamma((-\alpha, \alpha)) \subset M$, $\gamma(0) = \underline{a}$ and $\gamma'(0) = \sum_{j=1}^{k} y_j D_j G(\underline{a}_0)$. Notice also that for any $\underline{v} \in T_{\underline{a}} M$ we can choose these $y_1, \ldots, y_k$ such that the $\sum_{j=1}^{k} y_j D_j G(\underline{a}_0) = \underline{v}$ (by (5) in the above proof), i.e., γ can be selected so that $\gamma'(0) = \underline{v}$.

Given a C^1 map $f : U \to \mathbb{R}$, where U is an open subset of $\mathbb{R}^n$ with $U \supset M$, we define the *tangential gradient relative to M of f*, denoted $\nabla_M f$, (also referred to as the "gradient of $f|M$") by

9.6 $$\nabla_M f(\underline{a}) = P_{T_{\underline{a}} M}(\nabla_{\mathbb{R}^n} f(\underline{a})), \quad \underline{a} \in M \,,$$

where $P_{T_{\underline{a}} M} : \mathbb{R}^n \to \mathbb{R}^n$ is the orthogonal projection onto the tangent space $T_{\underline{a}} M$ (as in 8.3 of Ch. 1 with $V = T_{\underline{a}} M$). Recall that by 8.3 of Ch. 1 in fact the definition 9.6 simply says that $\nabla_M f(\underline{a})$ is the $T_{\underline{a}} M$-component $\underline{v}$ in the decomposition $\nabla_{\mathbb{R}^n} f(\underline{a}) = \underline{v} + \underline{u}$ with $\underline{v} \in T_{\underline{a}} M$ and $\underline{u} \in (T_{\underline{a}} M)^{\perp}$, hence the name "tangential gradient."

Of course since $P_{T_{\underline{a}} M}$ is a linear map we have that $\nabla_M f$ is also linear in f in the sense that if $f_j : U_j \to \mathbb{R}$ is C^1 with U_j open and $U_j \supset M$ for $j = 1, 2$, and if $\lambda, \mu \in \mathbb{R}$, then

9.7 $$\nabla_M(\lambda f_1 + \mu f_2)(\underline{a}) = \lambda \nabla_M f_1(\underline{a}) + \mu \nabla_M f_2(\underline{a}), \quad \underline{a} \in M \,.$$

9.8 Lemma. *Suppose M is a k-dimensional C^1 submanifold of $\mathbb{R}^n$ and $f : U \to \mathbb{R}$ is C^1, where U is an open set containing M. Then at each point $\underline{a} \in M$ we have*

(∗) $$\underline{v} \in T_{\underline{a}} M \Rightarrow D_{\underline{v}} f(\underline{a}) = \underline{v} \cdot \nabla_{\mathbb{R}^n} f(\underline{a}) = \underline{v} \cdot \nabla_M f(\underline{a}) \,.$$

Furthermore, if $f : \widetilde{U} \to \mathbb{R}$ is another C^1 function with $\widetilde{U} \supset M$ and if $\widetilde{f}|M = f|M$, then $\nabla_M f(\underline{a}) = \nabla_M \widetilde{f}(\underline{a}) \, \forall \underline{a} \in M$.

9.9 Remarks: (1) Notice, in particular, that (∗) implies that all the directional derivatives of f by vectors $\underline{v} \in T_{\underline{a}} M$ are determined by the tangential gradient $\nabla_M f(\underline{a})$.

(2) In the last part of the above lemma we use the notation $f|M$ to mean the "restriction of f to M," meaning the function which is defined on M only (and not on the larger set U) and which agrees with f at all points of M; the last part of the lemma in fact shows that the gradient of f on $M, \nabla_M f$, depends only on $f|M$ and is unchanged if we take another f, provided the new f is still C^1 and agrees with the original f at all points of M.

Proof of 9.8: For $\underline{v} \in T_{\underline{a}} M$ Def. 9.3 ensures that there is a C^1 map $\gamma : [0, \alpha) \to \mathbb{R}^n$ with $\alpha > 0$, $\gamma([0, \alpha)) \subset M$, $\gamma(0) = \underline{a}$, and $\gamma'(0) = \underline{v}$. By the chain rule, $\frac{d}{dt} f(\gamma(t)) = \gamma'(t) \cdot \nabla_{\mathbb{R}^n} f(\gamma(t))$ so, in particular, $\frac{d}{dt} f(\gamma(t))|_{t=0} = \underline{v} \cdot \nabla_{\mathbb{R}^n} f(\underline{a}) = D_{\underline{v}} f(\underline{a})$. On the other hand, since $\underline{v} \in T_{\underline{a}} M$ we have $\underline{v} = P_{T_{\underline{a}} M}(\underline{v})$ and then by 8.3 (with $V = T_{\underline{a}} M$) we have $\underline{v} \cdot \nabla_{\mathbb{R}^n} f(\underline{a}) = P_{T_{\underline{a}} M}(\underline{v}) \cdot \nabla_{\mathbb{R}^n} f(\underline{a}) = \underline{v} \cdot P_{T_{\underline{a}} M}(\nabla_{\mathbb{R}^n} f(\underline{a})) = \underline{v} \cdot \nabla_M f(\underline{a})$ by the definition 9.6.

To prove that last part of the lemma, observe that under the stated hypotheses $\widetilde{f} - f$ is C^1 on the open set $\widetilde{U} \cap U \supset M$, and so we can apply the above argument with $\widetilde{f} - f$ in place of f. Since $\widetilde{f} - f = 0$ at each point of M, this gives $\underline{v} \cdot \nabla_M(\widetilde{f} - f)(\underline{a}) = \underline{0} \, \forall \underline{a} \in M$ and all $\underline{v} \in T_{\underline{a}} M$ and so taking $\underline{v} = \nabla_M(\widetilde{f} - f)(\underline{a})$ we conclude $\nabla_M(\widetilde{f} - f)(\underline{a}) = \underline{0}$. Then by the linearity 9.7 this can be written $\nabla_M \widetilde{f} - \nabla_M f(\underline{a}) = \underline{0}$, so the proof of 9.8 is complete.

In view of the above theorem the following definition is natural.

9.10 Definition: Suppose M is a k-dimensional C^1 submanifold of $\mathbb{R}^n$ and $f : U \to \mathbb{R}$ is C^1, where U is an open set containing M. Then a point $\underline{a} \in M$ is called a critical point of $f|M$ (the restriction of f to M as in 9.9(2)) if $\nabla_M f(\underline{a}) = \underline{0}$.

Using the terminology of the above definition we then have the following.

9.11 Lemma. *If $f|M$ has a local maximum or minimum at $\underline{a} \in M$ (i.e., if there is $\delta > 0$ such that either $f(\underline{x}) \le f(\underline{a}) \,\forall\, \underline{x} \in M \cap B_\delta(\underline{a})$ or $f(\underline{x}) \ge f(\underline{a}) \,\forall\, \underline{x} \in M \cap B_\delta(\underline{a})$) then $\underline{a}$ is a critical point of $f|M$ (i.e., $\nabla_M f(\underline{a}) = \underline{0}$).*

9.12 Caution: Of course it need not be true under the hypotheses of the above lemma that the full $\mathbb{R}^n$ gradient $\nabla_{\mathbb{R}^n} f(\underline{a}) = \underline{0}$, because we are only assuming $f|M$ has a local maximum or minimum at $\underline{a}$, and this gives no information about the directional derivatives of f at $\underline{a}$ by vectors $\underline{v}$ which are not in $T_{\underline{a}} M$.

Proof of Lemma 9.11: Assume $f|M$ has a local maximum at $\underline{a}$, and take $\delta > 0$ such that $f(\underline{x}) \le f(\underline{a})$ for all $\underline{x} \in M \cap B_\delta(\underline{a})$. Let $\underline{v} \in T_{\underline{a}} M$. By Rem. 9.5 we can find a C^1 map $\gamma : (-\alpha, \alpha) \to \mathbb{R}^n$ with $\gamma((-\alpha, \alpha)) \subset M$, $\gamma(0) = 0$, and $\gamma'(0) = \underline{v}$. γ is C^1, hence continuous, and so there is an $\eta > 0$ such that $|t| < \eta \Rightarrow \|\gamma(t) - \gamma(0)\| < \delta$, so, since $\gamma(0) = \underline{a}$, we have $|t| < \eta \Rightarrow \gamma(t) \in M \cap B_\delta(\underline{a})$, and hence $|t| < \eta \Rightarrow f(\gamma(t)) \le f(\underline{a})$, so $f(\gamma(t))$ has a local maximum at $t = 0$. Hence, by 1-variable calculus the derivative at $t = 0$ is zero. That is, by the Chain Rule and 9.8, $0 = \frac{d}{dt} f(\gamma(t))|_{t=0} = \gamma'(0) \cdot \nabla_{\mathbb{R}^n} f(\underline{a}) = \underline{v} \cdot \nabla_{\mathbb{R}^n} f(\underline{a}) = \underline{v} \cdot \nabla_M f(\underline{a})$, so $\underline{v} \cdot \nabla_M f(\underline{a}) = 0$ for each $\underline{v} \in T_{\underline{a}} M$. In particular, with $\underline{v} = \nabla_M f(\underline{a})$ we infer $\|\nabla_M f(\underline{a})\|^2 = 0$, hence $\nabla_M f(\underline{a}) = \underline{0}$.

If $f|M$ has a local minimum at $\underline{a}$ the same argument can be applied to $-f$, hence the proof of 9.11 is complete.

One of the things we shall prove later (in Ch. 4) is that if $k \in \{1, \ldots, n-1\}$ then intersection of the level sets of the $n - k$ C^1 functions $g_1, \ldots, g_{n-k}$ is a k-dimensional C^1 submanifold, at least near points of the intersection where the gradients $\nabla_{\mathbb{R}^n} g_1, \ldots, \nabla_{\mathbb{R}^n} g_{n-k}$ are l.i. More precisely:

9.13 Theorem. *If $U \subset \mathbb{R}^n$ is open, if $g_1, \ldots, g_{n-k} : U \to \mathbb{R}$ are C^1, if $S = \{\underline{x} \in U : g_j(\underline{x}) = 0 \text{ for each } j = 1, \ldots, n - k\}$ and if $M \ne \emptyset$, where*

$$M = \{\underline{y} \in S : \nabla_{\mathbb{R}^n} g_1(\underline{y}), \ldots, \nabla_{\mathbb{R}^n} g_{n-k}(\underline{y}) \text{ are l.i.}\} ,$$

then M is a k-dimensional C^1 submanifold.

We shall give the proof of this later, in Sec. 3 of Ch. 4, as a corollary of the Implicit Function Theorem.

9.14 Lagrange Multiplier Theorem. *If $U \subset \mathbb{R}^n$ is open, if $f, g_1, \ldots, g_{n-k} : U \to \mathbb{R}$ are C^1, if $S = \{\underline{x} \in U : g_j(\underline{x}) = 0 \text{ for each } j = 1, \ldots, n - k\}$, if $\underline{a} \in M \equiv \{\underline{y} \in S : \nabla g_1(\underline{y}), \ldots,$*

$\nabla g_{n-k}(\underline{y})$ are l.i.} *(so that M is a k-dimensional C^1 manifold by 9.13), and if $\underline{a}$ is a critical point of $f|M$, then $\exists \lambda_1, \ldots, \lambda_{n-k} \in \mathbb{R}$ such that $\nabla_{\mathbb{R}^n} f(\underline{a}) = \sum_{j=1}^{n-k} \lambda_j \nabla_{\mathbb{R}^n} g_j(\underline{a})$.*

9.15 Terminology: The constants $\lambda_1, \ldots, \lambda_{n-k}$ are called *Lagrange Multipliers for f relative to the constraints* $g_1, \ldots, g_{n-k}$.

9.16 Remark: In view of 9.11 the hypothesis that $\underline{a}$ is a critical point of $f|M$ is automatically true (and hence the conclusion of 9.14 applies) if $f|M$ has a local maximum or local minimum at $\underline{a}$ (or, in particular, if $f|S$ has a local maximum or local minimum at $\underline{a} \in M$).

Proof of 9.14: By definition of critical point of $f|M$ we have

$$P_{T_{\underline{a}} M}(\nabla_{\mathbb{R}^n} f(\underline{a})) = \nabla_M f(\underline{a}) = \underline{0},$$

and hence $\nabla_{\mathbb{R}^n} f(\underline{a}) \in (T_{\underline{a}} M)^{\perp}$. By 9.4, $T_{\underline{a}} M$ has dimension k, and hence, by 8.1(iii) of Ch. 1, we have $\dim(T_{\underline{a}} M)^{\perp} = n - k$.

Now $g_j \equiv 0$ on M, hence by Lem. 9.8 above we have $\nabla_M g_j \equiv 0$ on M and hence we also have $\nabla_{\mathbb{R}^n} g_j(\underline{a}) \in (T_{\underline{a}} M)^{\perp}$. Furthermore, $\nabla_{\mathbb{R}^n} g_1(\underline{a}), \ldots, \nabla_{\mathbb{R}^n} g_{n-k}(\underline{a})$ are l.i., and hence, by Lem. 5.7(b) of Ch. 1, they must be a basis for $(T_{\underline{a}} M)^{\perp}$, so there are unique $\lambda_1, \ldots, \lambda_{n-k} \in \mathbb{R}$ such that $\nabla_{\mathbb{R}^n} f(\underline{a}) = \sum_{j=1}^{n-k} \lambda_j \nabla_{\mathbb{R}^n} g_j(\underline{a})$.

SECTION 9 EXERCISES

9.1 (a) Find the maximum value of the product $x_1 x_2 \cdots x_n$ subject to the constraints $x_j > 0$ for each $j = 1, \ldots, n$ and $\sum_{j=1}^{n} x_j = 1$, and (rigorously) justify your answer.

(b) Prove that if $a_1, \ldots, a_n$ are positive numbers, then $(a_1 a_2 \cdots a_n)^{1/n} \le \frac{a_1 + a_2 + \cdots + a_n}{n}$. ("The inequality between the arithmetic and geometric mean.")

9.2 Find:

(a) the minimum value of $x^2 + y^2$ on the line $x + y = 2$;

(b) the maximum value of $x + y$ on the circle $x^2 + y^2 = 2$.

9.3 Suppose $f(x, y, z) = 3xy + z^3 - 3z$ for $(x, y, z) \in \mathbb{R}^3$.

(i) Show that f has no local max or min in $\mathbb{R}^3$.

(ii) Prove that f attains its max and min values on the sphere $x^2 + y^2 + z^2 - 2z = 0$.

(iii) Find all the points on the sphere $x^2 + y^2 + z^2 - 2z = 0$ where f attains its max and min values.

9.4 Let $n \ge 2$ and $S^{n-1} = \{\underline{x} \in \mathbb{R}^n : \|\underline{x}\| = 1\}$. Prove that S^{n-1} is an $(n-1)$-dimensional C^1 submanifold.

Hint: Notice that we can write $S^{n-1} = \left(\cup_{j=1}^{n} \left\{ \underline{x} \in \mathbb{R}^n : \sum_{i \ne j} x_i^2 < 1 \text{ and } x_j = \sqrt{1 - \sum_{i \ne j} x_i^2} \right\} \right)$
$\cup \left(\cup_{j=1}^{n} \left\{ \underline{x} \in \mathbb{R}^n : \sum_{i \ne j} x_i^2 < 1 \text{ and } x_j = -\sqrt{1 - \sum_{i \ne j} x_i^2} \right\} \right)$.

CHAPTER 3

More Linear Algebra

1   PERMUTATIONS

A permutation on n elements is an ordered n-tuple $(i_1, \ldots, i_n)$ which is a re-ordering of the first n positive integers $1, \ldots, n$; that is the set $\mathcal{P}_n$ of all such permutations is

$$\mathcal{P}_n = \left\{ (i_1, \ldots, i_n) : \{i_1, \ldots, i_n\} = \{1, \ldots, n\} \right\}.$$

Notice that there are exactly $n!$ such permutations, because there are exactly n ways to choose the first entry i_1, and for each of these n choices there are exactly $n-1$ ways of choosing the entry i_2, so the number of ways of choosing the first two entries i_1, i_2 is $n(n-1)$, and similarly (if $n \geq 3$), a total of $n(n-1)(n-2)$ ways of choosing the first 3 entries i_1, i_2, i_3, and so on.

Let $1 \leq k < \ell \leq n$, and define $\tau_{k,\ell} : \mathbb{R}^n \to \mathbb{R}^n$ by

$$\tau_{k,\ell}(x_1, \ldots, \overset{\underset{\displaystyle\downarrow}{k\text{-th}}}{x_k}, \ldots, \overset{\underset{\displaystyle\downarrow}{\ell\text{-th}}}{x_\ell}, \ldots, x_n) = (x_1, \ldots, \overset{\underset{\displaystyle\downarrow}{k\text{-th}}}{x_\ell}, \ldots, \overset{\underset{\displaystyle\downarrow}{\ell\text{-th}}}{x_k}, \ldots, x_n)$$

(i.e., $\tau_{k,\ell}$ interchanges the elements at position k, ℓ). Such a transformation is called *a transposition*; notice that $\tau_{k,\ell}$ is its own inverse. Thus, $\tau_{k,\ell} \circ \tau_{k,\ell} = $ identity transformation on $\mathbb{R}^n$:

$$\tau_{k,\ell} \circ \tau_{k,\ell}(x_1, \ldots, x_n) \equiv (x_1, \ldots, x_n) \ \forall \, (x_1, \ldots, x_n) \in \mathbb{R}^n.$$

We claim that any permutation can be achieved by applying a finite sequence of such transpositions. Indeed, if $(i_1, \ldots, i_n)$ is any permutation then we can write this permutation in terms of a permutation of the first $n-1$ (instead of n) integers as follows:

$$(i_1, \ldots, i_n) = \begin{cases} \overbrace{(i_1, \ldots, i_{n-1}, n)}^{\text{perm. of } 1, \ldots, n-1,} & \text{if } i_n = n \\ \tau_{k,n}(\underbrace{i_1, \ldots, i_n, \ldots, i_{n-1}}_{k^{\text{th}}\text{slot}}, n) & \text{if } i_n < n \text{ (i.e., } i_k = n \text{ for some } k < n), \\ \underbrace{\phantom{\tau_{k,n}(i_1, \ldots, i_n, \ldots, i_{n-1}, n)}}_{\text{perm. of } 1, \ldots, n-1} \end{cases}$$

and so by induction on n it follows that each permutation can indeed be achieved by a sequence of transpositions; that is, for any permutation $(i_1, \ldots, i_n)$ we can find a sequence $\tau^{(1)}, \ldots, \tau^{(q)}$ of transpositions such that

1.1
$$(i_1, \ldots, i_n) = \tau^{(1)} \circ \tau^{(2)} \circ \cdots \circ \tau^{(q)}(1, \ldots, n).$$

Correspondingly, we can permute the n coordinates $x_1, \ldots, x_n$ of a point in $\mathbb{R}^n$ using the same sequence of transpositions:

1.2
$$(x_{i_1}, \ldots, x_{i_n}) = \tau^{(1)} \circ \tau^{(2)} \circ \cdots \circ \tau^{(q)}(x_1, x_2, \ldots, x_n) \, .$$

1.3 Note: We of course do not claim the representation 1.1 is unique, and it is not—indeed there are many different ways of writing a given permutation in terms of a sequence of transpositions. For example, if $n = 3$ then the permutation $(3, 2, 1)$ can be written as $\tau_{1,3}(1, 2, 3)$ but also as $\tau_{2,3} \circ \tau_{1,2} \circ \tau_{2,3}(1, 2, 3)$.

An important quantity related to a permutation $(i_1, \ldots, i_n)$ is the integer $N(i_1, \ldots, i_n)$, which is defined as the *number of inversions* present in the sequence $i_1, \ldots, i_n$; that is,

$$N(i_1, \ldots, i_n) = \text{ the number of pairs } k, \ell \in \{1, \ldots, n\} \text{ with } \ell > k \text{ and } i_\ell < i_k \, .$$

One usually computes this as $\sum_{j=1}^{n-1} N_j(i_1, \ldots, i_n)$, where

$$N_1(i_1, \ldots, i_n) = \text{number of integers } \ell > 1 \text{ such that } i_\ell < i_1$$
$$N_2(i_1, \ldots, i_n) = \text{the number of integers } \ell > 2 \text{ such that } i_\ell < i_2$$

$$\vdots \qquad\qquad \vdots$$

$$N_{n-1}(i_1, \ldots, i_n) = \text{the number of integers } \ell > n - 1 \text{ such that } i_\ell < i_{n-1}$$

(so that $N_{n-1}(i_1, \ldots, i_n) = 1$ if $i_n < i_{n-1}$ and $= 0$ if $i_n > i_{n-1}$).

The quantity $N(i_1, \ldots, i_n)$ allows us to define the "parity" (i.e., "evenness" or "oddness") of a permutation.

1.4 Definition: The permutation $(i_1, \ldots, i_n)$ is said to be even or odd according as the integer $N(i_1, \ldots, i_n)$ is even or odd.

Notice that then

$$(-1)^{N(i_1, \ldots, i_n)} = \begin{cases} +1 & \text{if } (i_1, \ldots, i_n) \text{ is even} \\ -1 & \text{if } (i_1, \ldots, i_n) \text{ is odd} \, . \end{cases}$$

The quantity $(-1)^{N(i_1, \ldots, i_n)}$ is called the *index* of the permutation $(i_1, \ldots, i_n)$.

1.5 Lemma. *If $1 \le k < \ell \le n$ and if $(i_1, \ldots, i_n)$ is a permutation of $(1, \ldots, n)$, then*

$$(-1)^{N(i_1, \ldots, i_\ell, \ldots, i_k, \ldots, i_n)} = -(-1)^{N(i_1, \ldots, i_k, \ldots, i_\ell, \ldots, i_n)} \, .$$

1.6 Remarks: (1) Notice that $(i_1, \ldots, i_\ell, \ldots, i_k, \ldots, i_n)$ can be written in terms of the transposition $\tau_{k,\ell}$ as $\tau_{k,\ell}(i_1, \ldots, i_k, \ldots, i_\ell, \ldots, i_n)$, so the above lemma says that *the parity of the permutation changes each time we apply a transposition*; notice, in particular, this means that the index $(-1)^{N(i_1, \ldots, i_n)}$ can be written alternatively as

$$(-1)^{N(i_1, \ldots, i_n)} = (-1)^q \, ,$$

where q is the number of transpositions appearing in 1.1 above.

(2) An additional consequence of the lemma is that for each $n \geq 2$ there are $\frac{n!}{2}$ even permutations and $\frac{n!}{2}$ odd permutations, because according to the lemma the transposition $\tau_{1,2}$ maps the even permutations to the odd permutations, and this map is clearly 1:1 and onto (it is onto because *any* permutation $(i_1, i_2, i_3, \ldots, i_n)$ can be written $\tau_{1,2}(i_2, i_1, i_3, \ldots, i_n)$).

Proof of Lemma 1.5: If $\ell = k + 1$, then i_k, i_ℓ are adjacent elements in the permutation $(i_1, \ldots, i_n)$. Observe that, if we use the notation that $\#A =$ the number of elements in the set A (zero if A is empty), then since the pair in the k-th and $(k + 1)$'st slots is i_{k+1}, i_k (which is an inversion if $i_{k+1} > i_k$ but not if $i_k < i_{k+1}$) we have

$$
\begin{aligned}
&\overset{k\text{-th}}{\downarrow} \qquad \overset{(k+1)\text{'st}}{\downarrow} \\
i_{k+1} > i_k \Rightarrow N_k(i_1, \ldots, i_{k-1}, i_{k+1},\ i_k, \ldots, i_n) &= 1 + \#\{j : j \geq k + 2 \text{ and } i_j < i_{k+1}\} \\
\text{and } N_{k+1}(i_1, \ldots, i_{k-1}, i_{k+1},\ i_k, \ldots, i_n) &= \quad\ \#\{j : j \geq k + 2 \text{ and } i_j < i_k\},
\end{aligned}
$$

so that

$$
\begin{aligned}
N_k(i_1, \ldots, i_{k-1},\ i_{k+1}, i_k, \ldots, i_n) + N_{k+1}(i_1, \ldots, i_{k-1},\ i_{k+1}, i_k, \ldots, i_n) = \\
1 + \#\{j : j \geq k + 2 \text{ and either } i_j < i_{k+1} \text{ or } i_j < i_k\},
\end{aligned}
$$

which, except for the "1+" on the right side, is exactly the same as

$$
N_k(i_1, \ldots, i_{k-1},\ i_k, i_{k+1}, \ldots, i_n) + N_{k+1}(i_1, \ldots, i_{k-1},\ i_k, i_{k+1}, \ldots, i_n).
$$

That is, we have shown that

$$
\begin{aligned}
i_{k+1} > i_k \Rightarrow &N_k(i_1, \ldots, i_{k-1},\ i_{k+1}, i_k, \ldots, i_n) + N_{k+1}(i_1, \ldots, i_{k-1},\ i_{k+1}, i_k, \ldots, i_n) = \\
&N_k(i_1, \ldots, i_{k-1},\ i_k, i_{k+1}, \ldots, i_n) + N_{k+1}(i_1, \ldots, i_{k-1},\ i_k, i_{k+1}, \ldots, i_n) + 1.
\end{aligned}
$$

By a similar argument (indeed it's the same identity applied to the permutation $(j_1, \ldots, j_n)$, where $j_\ell = i_\ell$ for $\ell \neq k, k + 1$ and $j_k = i_{k+1}$, $j_{k+1} = i_k$) we have

$$
\begin{aligned}
i_{k+1} < i_k \Rightarrow &N_k(i_1, \ldots, i_{k-1},\ i_{k+1}, i_k, \ldots, i_n) + N_{k+1}(i_1, \ldots, i_{k-1},\ i_{k+1}, i_k, \ldots, i_n) = \\
&N_k(i_1, \ldots, i_{k-1},\ i_k, i_{k+1}, \ldots, i_n) + N_{k+1}(i_1, \ldots, i_{k-1},\ i_k, i_{k+1}, \ldots, i_n) - 1.
\end{aligned}
$$

Since trivially $N_p(i_1, \ldots, i_{k-1}, i_{k+1}, i_k, \ldots, i_n) = N_p(i_1, \ldots, i_{k-1}, i_k, i_{k+1}, \ldots, i_n)$ if either $p < k$ or $p > k + 1$, then we have actually shown that

$$
N(i_1, \ldots, i_{k-1}, i_{k+1}, i_k, \ldots, i_n) = N(i_1, \ldots, i_{k-1}, i_k, i_{k+1}, \ldots, i_n) \pm 1
$$

and so $(-1)^{N(i_1, \ldots, i_{k-1}, i_{k+1}, i_k, \ldots, i_n)} = -(-1)^{N(i_1, \ldots, i_{k-1}, i_k, i_{k+1}, \ldots, i_n)}$ and the lemma is proved in the case when $\ell = k + 1$.

In the case when $\ell \geq k + 2$ we can get the permutation $(i_1, \ldots, i_\ell, \ldots, i_k, \ldots, i_n)$ from the original permutation $(i_1, \ldots, i_k, \ldots, i_\ell, \ldots, i_n)$ by first making a sequence of $\ell - k$ transpositions of adjacent elements to move i_k to the right of i_ℓ, and then making a sequence of $\ell - k - 1$ transpositions of adjacent elements to move i_ℓ back to the kth position. In all that is $2(\ell - k) - 1$ transpositions of adjacent elements, each changing the index by a factor of -1, so the total change in the index is $(-1)^{2(\ell-k)-1} = -1$, and the lemma is proved.

If $(i_1, \ldots, i_n)$ is any permutation of $1, \ldots, n$, recall that then we can pick transpositions $\tau^{(1)}, \ldots, \tau^{(q)}$ such that $(i_1, \ldots, i_n) = \tau^{(1)} \circ \tau^{(2)} \circ \cdots \circ \tau^{(q)}(1, \ldots, n)$. Then (since $\tau^{(i)} \circ \tau^{(i)} = $ identity for each i) the same sequence of transpositions in reverse order "undoes" this permutation: $(1, \ldots, n) = \tau^{(q)} \circ \cdots \circ \tau^{(1)}(i_1, \ldots, i_n)$. The permutation $(j_1, \ldots, j_n) = \tau^{(q)} \circ \cdots \circ \tau^{(1)}(1, \ldots, n)$ is called the "inverse" of $(i_1, \ldots, i_n)$. Observe that j_k is uniquely determined by the condition $i_{j_k} = k$ for each $k = 1, \ldots, n$, as one sees by taking $x_k = i_k$ in the identity $(x_{j_1}, \ldots, x_{j_n}) = \tau^{(q)} \circ \cdots \circ \tau^{(1)}(x_1, \ldots, x_n)$. Observe also that $(-1)^{N(j_1, \ldots, j_n)} = (-1)^q = (-1)^{N(i_1, \ldots, i_n)}$; that is, the permutation $(i_1, \ldots, i_n)$ and its inverse $(j_1, \ldots, j_n)$ have the same index.

SECTION 1 EXERCISES

1.1 For each of the following permutations, find the parity (i.e., evenness or oddness) by two methods: (i) by directly calculating the number N of inversions; and (ii) by representing the permutation by a sequence of transpositions applied to the trivial permutation $(1, 2, \ldots, n)$:

$$\text{(a) } (4, 3, 1, 5, 7, 2, 6), \quad \text{(b) } (2, 3, 1, 8, 5, 4, 7, 6).$$

2   DETERMINANTS

We here take the general "multilinear algebra" approach, which begins with the following abstract theorem (which in the course of the proof will become quite concrete).

2.1 Theorem. *There is one and only one function* $\mathcal{D} : \overbrace{\mathbb{R}^n \times \cdots \times \mathbb{R}^n}^{n\text{-factors}} \to \mathbb{R}$ *having the following properties:*

(i) *$\mathcal{D}$ is linear in the j-th factor for each $j = 1, \ldots, n$; For $j = 1$ this means*

$$\mathcal{D}(c_1\underline{\beta}_1 + c_2\underline{\gamma}_1, \underline{\alpha}_2, \ldots, \underline{\alpha}_n) = c_1\mathcal{D}(\underline{\beta}_1, \underline{\alpha}_2, \ldots, \underline{\alpha}_n) + c_2\mathcal{D}(\underline{\gamma}_1, \underline{\alpha}_2, \ldots, \underline{\alpha}_n),$$

for any $c_1, c_2 \in \mathbb{R}$ and any $\underline{\beta}_1, \underline{\gamma}_1, \underline{\alpha}_2, \ldots, \underline{\alpha}_n \in \mathbb{R}^n$, and we require a similar identity for the other factors;

(ii) *$\mathcal{D}$ changes sign when we interchange any pair of factors: that is, for any $k < \ell$,*

$$\mathcal{D}(\underline{\alpha}_1, \ldots, \underline{\alpha}_k, \ldots, \underline{\alpha}_\ell, \ldots, \underline{\alpha}_n) = -\mathcal{D}(\underline{\alpha}_1, \ldots, \underline{\alpha}_\ell, \ldots, \underline{\alpha}_k, \ldots, \underline{\alpha}_n);$$

(iii) *$\mathcal{D}(\underline{e}_1, \ldots, \underline{e}_n) = 1$.*

2.2 Remark: Observe that property (ii) says, in particular, that $\mathcal{D}(\underline{\alpha}_1, \ldots, \underline{\alpha}_n) = 0$ in case any two of the vectors $\underline{\alpha}_i, \underline{\alpha}_j$ $(i \neq j)$ are equal, because in that case by interchanging $\underline{\alpha}_i$ and $\underline{\alpha}_j$ and using property (ii) we conclude that $\mathcal{D}(\underline{\alpha}_1, \ldots, \underline{\alpha}_n)$ must be the negative of itself, i.e., it must be zero.

Proof: The proof is fairly straightforward now that we have the basic properties of permutations at our disposal: We start by *assuming* such a $\mathcal{D}$ exists, and seeing what conclusions that enables us to draw. In fact, if such $\mathcal{D}$ exists, then we can write each vector $\underline{\alpha}_j$ in the expression $\mathcal{D}(\underline{\alpha}_1, \ldots, \underline{\alpha}_n)$ as a linear combination of the standard basis vectors and use the linearity property (i). Thus, we write $\underline{\alpha}_1 = \sum_{i_1=1}^n a_{i_1 1}\underline{e}_{i_1}, \underline{\alpha}_2 = \sum_{i_2=1}^n a_{i_2 2}\underline{e}_{i_2}, \ldots, \underline{\alpha}_n = \sum_{i_n=1}^n a_{i_n n}\underline{e}_{i_n}$, and use the linearity property (i) on each of the sums. This gives the expression

$$\mathcal{D}(\underline{\alpha}_1, \ldots, \underline{\alpha}_n) = \sum_{i_1,i_2,\ldots,i_n=1}^n a_{i_1 1} a_{i_2 2} \cdots a_{i_n n} \mathcal{D}(\underline{e}_{i_1}, \underline{e}_{i_2}, \ldots, \underline{e}_{i_n}) .$$

In view of Rem. 2.2 above we can drop all terms where $i_\ell = i_k$ for some $k \neq \ell$; that is we are summing only over *distinct* $i_1, \ldots, i_n$, i.e., over $i_1, \ldots, i_n$ such that $(i_1, \ldots, i_n)$ is a permutation of the integers $1, \ldots, n$, so only $n!$ terms (from the total of n^n terms) are possibly nonzero, and we can write

$$\mathcal{D}(\underline{\alpha}_1, \ldots, \underline{\alpha}_n) = \sum_{\text{distinct } i_1,\ldots,i_n \leq n} a_{i_1 1} a_{i_2 2} \cdots a_{i_n n} \mathcal{D}(\underline{e}_{i_1}, \underline{e}_{i_2}, \ldots, \underline{e}_{i_n}) .$$

Now interchanging any two of the indices i_k, i_ℓ (with $k \neq \ell$) just changes the sign of the expression $\mathcal{D}(\underline{e}_{i_1}, \underline{e}_{i_2}, \ldots, \underline{e}_{i_n})$ (by virtue of property (ii)). So pick a sequence of transpositions $\tau^{(1)}, \ldots, \tau^{(q)}$ such that $(i_1, \ldots, i_n) = \tau^{(1)} \circ \tau^{(2)} \circ \cdots \circ \tau^{(q)}(1, \ldots, n)$; then $\mathcal{D}(e_{i_1}, \ldots, e_{i_n}) = (-1)^q \mathcal{D}(e_1, \ldots, e_n)$, and from our discussion of permutations we know that $(-1)^q$ is just the index $(-1)^{N(i_1,\ldots,i_n)}$ of the permutation $(i_1, \ldots, i_n)$, and also $\mathcal{D}(e_1, \ldots, e_n) = 1$ by property (iii), so we have proved (assuming a $\mathcal{D}$ satisfying (i), (ii), (iii) exists) that

$$(1) \qquad \mathcal{D}(\underline{\alpha}_1, \ldots, \underline{\alpha}_n) = \sum_{\text{perms. } (i_1,\ldots,i_n) \text{ of } \{1,\ldots,n\}} (-1)^{N(i_1,\ldots,i_n)} a_{i_1 1} a_{i_2 2} \cdots a_{i_n n} .$$

Thus, if such a map $\mathcal{D}$ exists, then there can only be one possible expression for it, Viz. the expression on the right of (1). Thus, to complete the proof we just have to show that if we *define* $\mathcal{D}(\underline{\alpha}_1, \ldots, \underline{\alpha}_n)$ to be the expression on the right side of (1), then $\mathcal{D}$ has the required properties (i), (ii), (iii).

Now properties (i), (iii) are obviously satisfied, so all we really need to do is check property (ii).

To check property (ii) observe first that i_k, i_ℓ are dummy variables so we can relabel them p, q, in which case

$$\mathcal{D}(\underline{\alpha}_1, .., \underset{\ell\text{-th}}{\underline{\alpha}_\ell}, .., \underset{k\text{-th}}{\underline{\alpha}_k}, .., \underline{\alpha}_n)$$

$$= \sum_{\text{distinct } i_1,\ldots p,\ldots,q,\ldots,i_n} (-1)^{N(i_1,\ldots,p,\ldots,q,\ldots,i_n)} a_{i_1 1} a_{i_2 2} \cdots \underset{k\text{-th}}{a_{p\ell}} \cdots \underset{\ell\text{-th}}{a_{qk}} \cdots a_{i_n n}$$

$$= \sum_{\text{distinct } i_1,\ldots p,\ldots,q,\ldots,i_n} (-1)^{N(i_1,\ldots,p,\ldots,q,\ldots,i_n)} a_{i_1 1} a_{i_2 2} \cdots a_{qk} \cdots a_{p\ell} \cdots a_{i_n n}$$

$$= - \sum_{\text{distinct } i_1,\ldots q,\ldots,p,\ldots,i_n} (-1)^{N(i_1,\ldots,q,\ldots,p,\ldots,i_n)} a_{i_1 1} a_{i_2 2} \cdots a_{qk} \cdots a_{p\ell} \cdots a_{i_n n} ,$$

where we used the fact that $(-1)^{N(i_1,\ldots,p,\ldots,q,\ldots,i_n)} = -(-1)^{N(i_1,\ldots,q,\ldots p,\ldots,i_n)}$ (by the main lemma, Lem. 1.5, in the previous section), and now we simply relabel $q = i_k$ and $p = i_\ell$ to see that the final expression here is indeed, as claimed, the quantity $-\mathcal{D}(\underline{\alpha}_1,\ldots,\underline{\alpha}_k,\ldots,\underline{\alpha}_\ell,\ldots,\underline{\alpha}_n)$.

Thus, the proof of the theorem is complete.

We can now define the determinant $\det A$ of an $n \times n$ matrix $A = (a_{ij})$.

2.3 Definition: $\det A = \mathcal{D}(\underline{\alpha}_1,\ldots,\underline{\alpha}_n)$, where $\underline{\alpha}_1,\ldots,\underline{\alpha}_n$ are the columns of A.

We can now check the various properties of determinant: The first 3 are just restatements of properties (i), (ii), (iii) of the theorem, and hence require no proof.

P1. $\det A$ is a linear function of each of the columns of A (assuming the other columns are held fixed).

P2. If we interchange 2 (different) columns of A, then the value of the determinant is changed by a factor -1 (and hence $\det A = 0$ if A has two equal columns).

P3. If $A = I$ (the identity matrix), then $\det A = 1$.

P4. If A, B are $n \times n$ matrices, then $\det AB = \det A \det B$.

Proof of P4: Let $A = (a_{ij})$, $B = (b_{ij})$. Recall that by definition of matrix multiplication we have that the j-th column of AB is $\sum_{k=1}^{n} b_{kj}\underline{\alpha}_k$, where $\underline{\alpha}_k$ is the k-th column of A. Thus, using the linearity and alternating properties (i), (ii) for $\mathcal{D}$ we have

$$\det AB = \mathcal{D}(\sum_{k_1=1}^{n} b_{k_11}\underline{\alpha}_{k_1},\ldots,\sum_{k_n=1}^{n} b_{k_nn}\underline{\alpha}_{k_n})$$

$$= \sum_{k_1,k_2,\ldots,k_n=1}^{n} b_{k_11}b_{k_22}\cdots b_{k_nn}\, \mathcal{D}(\underline{\alpha}_{k_1},\ldots,\underline{\alpha}_{k_n})$$

$$= \sum_{\text{distinct } k_1,k_2,\ldots,k_n} b_{k_11}b_{k_22}\cdots b_{k_nn}\, \mathcal{D}(\underline{\alpha}_{k_1},\ldots,\underline{\alpha}_{k_n})$$

$$= \sum_{\text{distinct } k_1,k_2,\ldots,k_n} (-1)^{N(k_1,\ldots,k_n)} b_{k_11}b_{k_22}\cdots b_{k_nn}\, \mathcal{D}(\underline{\alpha}_1,\ldots,\underline{\alpha}_n)$$

$$= \det B\ \det A$$

as claimed.

P5. $\det A = \det A^{\mathrm{T}}$ (and hence the properties **P1, P2** apply also to the *rows* of A).

P6. (Expansion along the i-th row or j-th column.) The following formulae are valid:

$$\det A = \sum_{j=1}^{n}(-1)^{i+j}a_{ij}\det A_{ij} \text{ for each } i = 1,\ldots,n$$
$$\det A = \sum_{i=1}^{n}(-1)^{i+j}a_{ij}\det A_{ij} \text{ for each } j = 1,\ldots,n\ ,$$

where A_{ij} denotes the $(n-1) \times (n-1)$ matrix obtained by deleting the i-th row and j-th column of the matrix A.

We defer the proof of **P5, P6** for a moment, in order to make an important point about practical computation of determinants:

2.4 Remark: (Highly relevant for the efficient computation of determinants.) Notice that one almost never directly uses the definition (or for that matter **P6**) to compute a determinant. Rather, one uses the properties **P1, P2** (and the corresponding facts about rows instead of columns which we can do by **P5**) in order to first drastically simplify the matrix before attempting to expand the determinant. In particular, observe that the 3rd kind of elementary row operation does not change the value of the determinant at all:

$$\det \begin{pmatrix} \underline{r}_1 \\ \underline{r}_2 \\ \vdots \\ \underline{r}_i \\ \vdots \\ \underline{r}_j + \lambda \underline{r}_i \\ \vdots \\ \underline{r}_n \end{pmatrix} = \det \begin{pmatrix} \underline{r}_1 \\ \underline{r}_2 \\ \vdots \\ \underline{r}_i \\ \vdots \\ \underline{r}_j \\ \vdots \\ \underline{r}_n \end{pmatrix} + \lambda \det \begin{pmatrix} \underline{r}_1 \\ \underline{r}_2 \\ \vdots \\ \underline{r}_i \quad \leftarrow i\text{-th} \\ \vdots \\ \underline{r}_i \quad \leftarrow j\text{-th} \\ \vdots \\ \underline{r}_n \end{pmatrix} = \det \begin{pmatrix} \underline{r}_1 \\ \underline{r}_2 \\ \vdots \\ \underline{r}_i \\ \vdots \\ \underline{r}_j \\ \vdots \\ \underline{r}_n \end{pmatrix}$$

where we used **P1** and **P2**. We can of course also use the other two types of elementary row operations, so long as we keep track of the fact that interchanging 2 rows changes the sign of the determinant and multiplying a row by a constant λ has the effect of multiplying the determinant by λ.

We now want to check properties **P5, P6**.

Proof of P5: Recall that if $(i_1, \ldots, i_n)$ is any permutation of $1, \ldots, n$ then we can pick transpositions $\tau^{(1)}, \ldots, \tau^{(q)}$ such that $(i_1, \ldots, i_n) = \tau^{(1)} \circ \tau^{(2)} \circ \cdots \circ \tau^{(q)}(1, \ldots, n)$, and then the same sequence of transpositions in reverse order "undoes" this permutation: $(1, \ldots, n) = \tau^{(q)} \circ \cdots \circ \tau^{(1)}(i_1, \ldots, i_n)$. The permutation $(j_1, \ldots, j_n) = \tau^{(q)} \circ \cdots \circ \tau^{(1)}(1, \ldots, n)$ is called the "inverse" of $(i_1, \ldots, i_n)$. Observe that j_k is uniquely determined by the condition $i_{j_k} = k$ for each $k = 1, \ldots, n$, as one sees by taking $x_k = i_k$ in the identity $(x_{j_1}, \ldots, x_{j_n}) = \tau^{(q)} \circ \cdots \circ \tau^{(1)}(x_1, \ldots, x_n)$. Observe also that $(-1)^{N(j_1, \ldots, j_n)} = (-1)^q = (-1)^{N(i_1, \ldots, i_n)}$; that is, the permutation $(i_1, \ldots, i_n)$ and its inverse $(j_1, \ldots, j_n)$ have the same index.

Now

$$\det A = \sum_{\text{distinct } i_1, \ldots, i_n} (-1)^{N(i_1, \ldots, i_n)} a_{i_1 1} a_{i_2 2} \cdots a_{i_n n},$$

and observe that the terms in the product $a_{i_1 1} a_{i_2 2} \cdots a_{i_n n}$ can be reordered so that the j_k-th entry (which is $a_{i_{j_k} j_k} = a_{k j_k}$) appears as the k-th factor. Thus,

$$a_{i_1 1} a_{i_2 2} \cdots a_{i_n n} = a_{1 j_1} a_{2 j_2} \cdots a_{n j_n} ,$$

and, as we mentioned above, the index $(-1)^{N(j_1,\ldots,j_n)}$ of $(j_1,\ldots,j_n)$ is the same as the index $(-1)^{N(i_1,\ldots,i_n)}$ of $(i_1,\ldots,i_n)$. We thus have

$$\det A = \sum_{\text{distinct } i_1,\ldots,i_n} (-1)^{N(j_1,\ldots,j_n)} a_{1 j_1} a_{2 j_2} \cdots a_{n j_n} .$$

Since the correspondence $(i_1,\ldots,i_n) \to (j_1,\ldots,j_n)$ between a permutation and its inverse is 1:1 (hence onto all $n!$ permutations), we see that $(j_1,\ldots,j_n)$ visits each of the $n!$ permutations exactly once in the above sum over distinct $i_1,\ldots,i_n$. That is, we have shown

$$\det A = \sum_{\text{distinct } j_1,\ldots,j_n} (-1)^{N(j_1,\ldots,j_n)} a_{1 j_1} a_{2 j_2} \cdots a_{n j_n} .$$

But now observe what we have here is exactly the expression for the determinant of A^{T}. Thus, $\det A = \det A^{\mathrm{T}}$ as claimed in **P5**.

Proof of P6: In view of **P5** we need only prove the second identity here, because the first follows by applying the second to the transpose matrix. Furthermore, we need only prove the first identity in the case $j = n$, because the case $j = n$ implies the result for $j < n$ by virtue of property **P2**. Let $\underline{\alpha}_1,\ldots,\underline{\alpha}_n$ be the columns of A. Since $\underline{\alpha}_n = \sum_{i=1}^{n} a_{in}\underline{e}_i$, we have

$$(1) \qquad \begin{aligned} \det A &= \mathcal{D}(\underline{\alpha}_1,\ldots,\underline{\alpha}_n) = \mathcal{D}(\underline{\alpha}_1,\ldots,\underline{\alpha}_{n-1}, \textstyle\sum_{i=1}^{n} a_{in}\underline{e}_i) \\ &= \sum_{i=1}^{n} a_{in} \mathcal{D}(\underline{\alpha}_1,\ldots,\underline{\alpha}_{n-1},\underline{e}_i) , \end{aligned}$$

and by making $n - i$ interchanges of adjacent rows, and using the fact that (by **P5**) property **P2** also applies with respect to interchanges of rows rather than columns, we can move the i-th row of the $n \times n$ matrix $(\underline{\alpha}_1\underline{\alpha}_2 \cdots \underline{\alpha}_{n-1}\underline{e}_i)$ to the n-th position, at the expense of gaining a factor $(-1)^{n-i}$; and so, since $(-1)^{n-i} = (-1)^{n+i}$, we obtain

$$(2) \qquad \mathcal{D}(\underline{\alpha}_1,\ldots,\underline{\alpha}_{n-1},\underline{e}_i) = (-1)^{n+i} \det \begin{pmatrix} A_{in} & \underline{0} \\ \underline{r}_i & 1 \end{pmatrix} ,$$

where $\underline{0}$ is the zero column vector in $\mathbb{R}^{n-1}$ and $\underline{r}_i$ is the row vector $(a_{i1}, a_{i2}, \ldots, a_{in-1}) \in \mathbb{R}^{n-1}$ (i.e., $\underline{r}_i$ is just the first $n - 1$ entries of the i-th row of A).

Now by definition

$$\det \begin{pmatrix} A_{in} & \underline{0} \\ \underline{r}_i & 1 \end{pmatrix} = \sum_{\text{distinct } i_1,\ldots,i_n \leq n} (-1)^{N(i_1,\ldots,i_n)} b_{i_1 1} b_{i_2 2} \ldots b_{i_n n} ,$$

where b_{ij} is the element in the i-th row and j-th column of the matrix $\begin{pmatrix} A_{in} & 0 \\ \underline{r}_i & 1 \end{pmatrix}$; notice that

$b_{i_n n} = 0$ if $i_n \neq n$ and 1 if $i_n = n$, so this expression simplifies to

$$\det \begin{pmatrix} A_{in} & 0 \\ \underline{r}_i & 1 \end{pmatrix} = \sum_{\text{distinct } i_1, \ldots, i_{n-1} \leq n-1} (-1)^{N(i_1, \ldots, i_{n-1}, n)} b_{i_1 1} b_{i_2 2} \ldots b_{i_{n-1} n-1} ,$$

which of course (since $N(i_1, \ldots, i_{n-1}, n) = N(i_1, \ldots, i_{n-1})$) is just exactly the expression for determinant of the $(n-1) \times (n-1)$ matrix obtained by deleting the n-th row and the n-th column of the matrix $\begin{pmatrix} A_{in} & 0 \\ \underline{r}_i & 1 \end{pmatrix}$, i.e., it is the determinant $\det A_{in}$; that is,

(3) $$\det \begin{pmatrix} A_{in} & 0 \\ \underline{r}_i & 1 \end{pmatrix} = \det A_{in} .$$

By combining (1), (2), (3) we have **P6** as claimed.

SECTION 2 EXERCISES

2.1 Calculate the determinant of

$$\begin{pmatrix} 10 & 11 & 12 & 13 & 426 \\ 2000 & 2001 & 2002 & 2003 & 421 \\ 2 & 2 & 1 & 0 & 419 \\ 100 & 101 & 101 & 102 & 2000 \\ 2003 & 2004 & 2005 & 2006 & 421 \end{pmatrix}$$

(show all row/column operations: no calculators).

2.2 Prove that the "Vandermonde determinant" $\Delta = \det \begin{pmatrix} 1 & 1 & \cdots & 1 \\ x_1 & x_2 & \cdots & x_n \\ x_1^2 & x_2^2 & \cdots & x_n^2 \\ \vdots & & & \vdots \\ x_1^{n-1} & x_2^{n-1} & \cdots & x_n^{n-1} \end{pmatrix}$

is the product of all possible factors $x_j - x_i$ with $1 \leq i < j \leq n$; that is, $\Delta = \Pi_{1 \leq i < j \leq n}(x_j - x_i)$.

Hint: Consider using row operations to reduce the n-variable case to the $(n-1)$-variable case.

3   INVERSE OF A SQUARE MATRIX

One very important theoretical consequence of the formula **P6** of the previous section is that it enables us to prove the *inverse* of an $n \times n$ matrix A (i.e., the $n \times n$ matrix A^{-1} such that $A^{-1}A = AA^{-1} = I$)

exists if $\det A \neq 0$. First observe that according to the first identity in **P6** we have

$$\sum_{k=1}^{n}(-1)^{i+k}a_{\ell k}\det A_{ik} = \det \begin{pmatrix} a_{11} & a_{12} & \cdots & a_{1n} \\ a_{21} & a_{22} & \cdots & a_{2n} \\ \vdots & \vdots & & \vdots \\ a_{\ell 1} & a_{\ell 2} & \cdots & a_{\ell n} & \leftarrow i\text{-th row} \\ \vdots & \vdots & & \vdots \\ a_{\ell 1} & a_{\ell 2} & \cdots & a_{\ell n} & \leftarrow \ell\text{-th row} \\ \vdots & \vdots & & \vdots \\ a_{n1} & a_{n2} & \cdots & a_{nn} \end{pmatrix}$$

which is zero if $i \neq \ell$. On the other hand, if $i = \ell$ then **P6** says the expression on the left is just $\det A$. Thus, we have

$$\sum_{k=1}^{n}(-1)^{i+k}a_{\ell k}\det A_{ik} = \begin{cases} \det A & \text{if } i = \ell \\ 0 & \text{if } i \neq \ell, \end{cases}$$

or in other words (labeling the indices differently)

$$\sum_{k=1}^{n}(-1)^{j+k}a_{ik}\det A_{jk} = \begin{cases} \det A & \text{if } i = j \\ 0 & \text{if } i \neq j. \end{cases}$$

Notice that the expression on the left is just the element in the i-th row and j-th column of the matrix product AM, where M is the $n \times n$ matrix with entry m_{ij} in the i-th row and j-th column, where $m_{ij} = (-1)^{i+j}\det A_{ji}$; in other words,

$$M = \left((-1)^{i+j}\det A_{ji}\right),$$

and then the above, in matrix terminology, just says

$$AM = (\det A)\, I,$$

where I is the $n \times n$ identity matrix. By using an almost identical argument, based on the second identity in **P6** rather than the first, we also deduce

$$MA = (\det A)\, I,$$

where M is the same matrix.

3.1 Terminology: The above matrix M (which has the entry $(-1)^{i+j}\det A_{ji}$ in its i-th row and j-th column) is called the "adjoint matrix of A," and is denoted adj A.

Because of the above we now have a complete understanding of when an $n \times n$ matrix A has an *inverse*, i.e., when there is an $n \times n$ matrix A^{-1} such that $A^{-1}A = AA^{-1} = I$.

3.2 Theorem. *Let A be an n × n matrix. Then A has an inverse if and only if* $\det A \neq 0$, *and in this case the inverse* A^{-1} *is given explicitly by* $A^{-1} = \frac{1}{\det A} \operatorname{adj} A$.

Proof: First assume $\det A \neq 0$. Then by the formulae $\operatorname{adj} A \, A = A \operatorname{adj} A = (\det A) \, I$ proved above, we see that indeed the inverse exists and is given explicitly by the formula $\frac{1}{\det A} \operatorname{adj} A$.

Conversely, if A^{-1} exists, then we have $A A^{-1} = I$, and hence taking determinants of each side and using properties **P4** and **P3** we obtain $(\det A)(\det A^{-1}) = 1$, which implies $\det A \neq 0$, so the proof is complete.

3.3 Remark: Notice that the last part of the above argument gives the additional information that

$$\det A^{-1} = \frac{1}{\det A} \, .$$

Recall (see the remark after the statement of properties **P1**–**P6**) that we can compute $\det A$ by using elementary row operations and that the first type of elementary row operation (interchanging any 2 rows) only changes the sign, the second type (multiplying a row by a nonzero scalar λ) multiplies the determinant by λ ($\neq 0$), and the third type (replacing the i-th row by the i-th row plus any multiple of the j-th row with $i \neq j$) does not change the value of the determinant at all. Thus, none of the elementary row operations can change whether or not the matrix has zero determinant. In particular, this means that $\det A \neq 0 \iff \det \operatorname{rref} A \neq 0$. On the other hand, since A is $n \times n$ (i.e., we are in the case $m = n$), then, from our previous discussion of $\operatorname{rref} A$ in Sec. 9. of Ch. 1, we know that either $\operatorname{rref} A = I$ (the $n \times n$ identity matrix) or $\operatorname{rref} A$ has a zero row (in which case it trivially has determinant $= 0$). We also recall (again from Sec. 9 of Ch. 1) that (since we are in the case $m = n$) $\operatorname{rref} A$ has a zero row if and only if the null space $N(A)$ is nontrivial (i.e., if and only if the homogeneous system $A\underline{x} = \underline{0}$ has a nontrivial solution), which in turn is true if and only if A has rank less than n.

Thus, combining these facts with Thm. 3.2 above, we have:

3.4 Theorem. *If A is an n × n matrix, then*

$$A^{-1} \text{ exists} \iff \det A \neq 0 \iff \operatorname{rref} A = I \iff N(A) = \{0\} \iff \operatorname{rank} A = n \, .$$

SECTION 3 EXERCISES

3.1 Suppose A, B are $n \times n$ matrices and $AB = I$. Prove that then $\det A \neq 0$ and $B = (\det A)^{-1} \operatorname{adj} A$. (In particular, $AB = I \Rightarrow BA = I$ and B is the unique inverse $(\det A)^{-1} \operatorname{adj} A$).

3.2 If A, B are, respectively, $m \times n$ and $n \times p$ matrices, prove

(a) $(AB)^{\mathsf{T}} = B^{\mathsf{T}} A^{\mathsf{T}}$, and (b) $m = n = p$ and A, B invertible $\Rightarrow AB$ invertible and $(AB)^{-1} = B^{-1} A^{-1}$.

3.3 Let $A = (a_{ij})$ be an $n \times n$ matrix with $\det A \neq 0$, and $\underline{b} = (b_1, \ldots, b_n)^{\mathrm{T}} \in \mathbb{R}^n$. Prove that the solution $\underline{x} = (x_1, \ldots, x_n)^{\mathrm{T}} = A^{-1}\underline{b}$ of the system $A\underline{x} = \underline{b}$ is given by the formula

$$x_i = \frac{\det A_i}{\det A}, \quad i = 1, \ldots, n,$$

where A_i is obtained from A by replacing the ith column of A by the given vector $\underline{b}$.

Hint: Use the formula $A^{-1} = (\det A)^{-1}\left((-1)^{i+j} \det A_{ji}\right)$ proved above to calculate the i-th component of $A^{-1}\underline{b}$.

4   COMPUTING THE INVERSE

By Exercise 3.1 above, an $n \times n$ matrix B is the inverse of the $n \times n$ matrix A if and only if $AB = I$, where I is the $n \times n$ identity matrix, i.e., the matrix with j-th column $\underline{e}_j$. Since the j-th column of AB is $A\underline{\beta}_j$, where $\underline{\beta}_j$ is the j-th column of B, to find the inverse of A we thus have to solve

4.1 $$A\underline{x} = \underline{e}_j, \quad j = 1, \ldots, n.$$

In view of the discussion in Sec. 10 of Ch. 1 we can attempt to do this by using elementary row operations on the augmented matrix $A|\underline{e}_j$ to give the new augmented matrix

4.2 $$\mathrm{rref}\, A|\underline{\beta}_j,$$

and by Thm. 3.4 above we know that A^{-1} exists if and only if $\mathrm{rref}\, A = I$, in which case the augmented system in 4.2 is simply

4.3 $$I|\underline{\beta}_j, j = 1, \ldots, n \ddot{y},$$

and hence if A^{-1} exists then the solution of 4.1, which gives the j-th column of A^{-1}, is exactly the vector which appears right side of 4.2 (and in that case $\mathrm{rref}\, A = I$). We can of course solve all j equations in 4.1 simultaneously by working with the $n \times 2n$ augmented matrix $A|I$ (because the j-th column of I is $\underline{e}_j$).

Thus, to summarize, we can find the inverse by starting with the $n \times 2n$ augmented matrix $A|I$ and applying elementary row operations, which produce $\mathrm{rref}\, A = I$ as the first n columns, and produces the $n \times 2n$ augmented matrix

$$I|B.$$

Then B is the required inverse A^{-1}. This process will also tell us when A^{-1} does not exist, because in that case $\mathrm{rref}\, A$ will have at least one row of zeros (and $\det A = 0$).

For example, if A is 3×3 matrix $\begin{pmatrix} 1 & 4 & 3 \\ 1 & 4 & 5 \\ 2 & 5 & 1 \end{pmatrix}$ then we compute the inverse (and at the same time check that the inverse exists) by looking at the 3×6 augmented matrix

$$\begin{pmatrix} 1 & 4 & 3 & 1 & 0 & 0 \\ 1 & 4 & 5 & 0 & 1 & 0 \\ 2 & 5 & 1 & 0 & 0 & 1 \end{pmatrix},$$

and using elementary row operations as follows:

$$\begin{pmatrix} 1 & 4 & 3 & | & 1 & 0 & 0 \\ 1 & 4 & 5 & | & 0 & 1 & 0 \\ 2 & 5 & 1 & | & 0 & 0 & 1 \end{pmatrix} \quad \begin{matrix} r_2 \to r_2 - r_1 \\ r_3 \to r_3 - 2r_1 \end{matrix}$$

$$\begin{pmatrix} 1 & 4 & 3 & | & 1 & 0 & 0 \\ 0 & 0 & 2 & | & -1 & 1 & 0 \\ 0 & -3 & -5 & | & -2 & 0 & 1 \end{pmatrix}$$

$$\begin{matrix} r_2 \to \frac{1}{2}r_2 \\ r_3 \to -r_3 \end{matrix}$$

$$\begin{pmatrix} 1 & 4 & 3 & | & 1 & 0 & 0 \\ 0 & 0 & 1 & | & -1/2 & 1/2 & 0 \\ 0 & 3 & 5 & | & 2 & 0 & -1 \end{pmatrix}$$

$$r_1 \to r_1 - 3r_2$$

$$\begin{pmatrix} 1 & 4 & 0 & | & 5/2 & -3/2 & 0 \\ 0 & 0 & 1 & | & -1/2 & 1/2 & 0 \\ 0 & 3 & 0 & | & 9/2 & -5/2 & -1 \end{pmatrix}$$

$$r_3 \to r_3 - 5r_2$$

$$\begin{pmatrix} 1 & 4 & 0 & | & 5/2 & -3/2 & 0 \\ 0 & 0 & 1 & | & -1/2 & 1/2 & 0 \\ 0 & 1 & 0 & | & 9/6 & -5/6 & -1/3 \end{pmatrix}$$

$$r_3 \to \frac{1}{3}r_3$$

$$r_1 \to r_1 - 4r_3$$

$$\begin{pmatrix} 1 & 0 & 0 & | & -21/6 & 11/6 & 4/3 \\ 0 & 0 & 1 & | & -1/2 & 1/2 & 0 \\ 0 & 1 & 0 & | & 9/6 & -5/6 & 1/3 \end{pmatrix}$$

$$r_2 \leftrightarrow r_3$$

$$\begin{pmatrix} 1 & 0 & 0 & | & -21/6 & 11/6 & 4/3 \\ 0 & 1 & 0 & | & 9/6 & -5/6 & 1/3 \\ 0 & 0 & 1 & | & -1/2 & 1/2 & 0 \end{pmatrix}$$

Therefore, $A^{-1} = \begin{pmatrix} -7/2 & 11/6 & 4/3 \\ 3/2 & -5/6 & -1/3 \\ -1/2 & 1/2 & 0 \end{pmatrix}$.

SECTION 4 EXERCISES

4.1 Calculate the inverse A^{-1} of A, if A is the 4×4 matrix $\begin{pmatrix} 1 & 1 & 1 & 1 \\ 0 & 1 & 1 & 1 \\ 1 & 0 & 2 & 3 \\ 0 & 0 & 1 & 2 \end{pmatrix}$.

5   ORTHONORMAL BASIS AND GRAM-SCHMIDT

Let $\underline{v}_1, \ldots, \underline{v}_N$ be vectors in $\mathbb{R}^n$. We say that the vectors are *orthonormal* if $\underline{v}_i \cdot \underline{v}_j = \delta_{ij}$, where δ_{ij} is the Kronecker delta ($= 1$ when $i = j$ and $= 0$ when $i \neq j$). Thus, $\underline{v}_1, \ldots, \underline{v}_N$ orthonormal means that each vector has length 1 and the vectors are mutually orthogonal (i.e., each vector is orthogonal to all the others).

If V is a nontrivial subspace of $\mathbb{R}^n$ then an *orthonormal basis* for V is a basis for V consisting of an orthonormal set of vectors.

5.1 Remark: Observe that orthonormal vectors $\underline{v}_1, \ldots, \underline{v}_N$ are automatically l.i., because

$$\sum_{i=1}^{N} c_i \underline{v}_i = \underline{0} \Rightarrow c_j = 0 \ \forall \ j = 1, \ldots, N \ ,$$

as we see by taking the dot product of each side of the identity $\sum_{i=1}^{N} c_i \underline{v}_i = \underline{0}$ with the vector $\underline{v}_j$.

5.2 Lemma. *If V is a nontrivial subspace of $\mathbb{R}^n$ of dimension k and if $\underline{u}_1, \ldots, \underline{u}_k$ is an orthonormal basis for V, then the orthogonal projection P_V of $\mathbb{R}^n$ onto V (as in Thm. 8.3 of Ch. 1) is given explicitly by*

$$P_V(\underline{x}) = \sum_{i=1}^{k} \underline{x} \cdot \underline{u}_i \, \underline{u}_i, \quad \underline{x} \in \mathbb{R}^n \ .$$

Proof: Define $Q(\underline{x}) = \sum_{i=1}^{k} \underline{x} \cdot \underline{u}_i \, \underline{u}_i$ for $\underline{x} \in \mathbb{R}^n$, and observe that then $Q(\underline{x}) \in V$ by definition. Also, since $\underline{u}_j \cdot \underline{u}_i = \delta_{ij}, \underline{u}_j \cdot (\underline{x} - Q(\underline{x})) = \underline{u}_j \cdot \underline{x} - \underline{x} \cdot \underline{u}_j = 0$, so that $\underline{x} - Q(\underline{x})$ is orthogonal to each $\underline{u}_j$, hence orthogonal to $\mathrm{span}\{\underline{u}_1, \ldots, \underline{u}_k\}$. That is, $\underline{x} \in V$ and $\underline{x} - Q(\underline{x}) \in V^\perp$ and hence $Q(\underline{x})$ satisfies the two conditions which according to Thm. 8.3 of Ch. 1 uniquely determine the orthogonal projection P_V. Hence, $Q = P_V$ and the proof is complete.

Our main result here is the following.

5.3 Theorem. *Let $\underline{v}_1, \ldots, \underline{v}_k$ be any l.i. set of vectors in $\mathbb{R}^n$. Then there is an orthonormal set of vectors $\underline{u}_1, \ldots, \underline{u}_k$ such that $\mathrm{span}\{\underline{u}_1, \ldots, \underline{u}_j\} = \mathrm{span}\{\underline{v}_1, \ldots, \underline{v}_j\}$ for each $j = 1, \ldots, k$, and in fact $\underline{u}_1, \ldots, \underline{u}_j$ is an orthonormal basis for $\mathrm{span}\{\underline{v}_1, \ldots, \underline{v}_j\}$ for each $j = 1, \ldots, k$.*

Proof ("Gram-Schmidt orthogonalization"): We proceed inductively to construct orthonormal $\underline{u}_1, \underline{u}_2, \ldots, \underline{u}_k$ such that $\mathrm{span}\{\underline{u}_1, \ldots, \underline{u}_j\} = \mathrm{span}\{\underline{v}_1, \ldots, \underline{v}_j\}$ for each j by the following procedure, in which we use the notation that P_j denotes the orthogonal projection of $\mathbb{R}^n$ onto $\mathrm{span}\{\underline{v}_1, \ldots, \underline{v}_j\}$ for each $j = 1, \ldots, k$:

$$\underline{u}_1 = \|\underline{v}_1\|^{-1} \underline{v}_1$$

and, assuming $j \in \{2, \ldots, k-1\}$ and that we have selected orthonormal $\underline{u}_1, \ldots, \underline{u}_j$ with $\mathrm{span}\{\underline{u}_1, \ldots, \underline{u}_j\} = \mathrm{span}\{\underline{v}_1, \ldots, \underline{v}_j\}$, we define

$$(1) \qquad \underline{u}_{j+1} = \|\underline{v}_{j+1} - P_j(\underline{v}_{j+1})\|^{-1} (\underline{v}_{j+1} - P_j(\underline{v}_{j+1})) \ .$$

Observe that this makes sense because $\underline{v}_{j+1} - P_j(\underline{v}_j) \neq \underline{0}$ (because by definition $P_j(\underline{v}_{j+1})$ is a linear combination of $\underline{v}_1, \ldots, \underline{v}_j$ and so $\underline{v}_{j+1} - P_j(\underline{v}_j)$ is a nontrivial linear combination of $\underline{v}_1, \ldots, \underline{v}_{j+1}$ and hence cannot be zero because $\underline{v}_1, \ldots, \underline{v}_{j+1}$ are l.i.).

Then by definition, $\underline{u}_{j+1}$ is a unit vector, and if $i = 1, \ldots, j$ then we have

$$\underline{u}_i \cdot \underline{u}_{j+1} = c(\underline{u}_i \cdot \underline{v}_{j+1} - \underline{u}_i \cdot P_j(\underline{v}_{j+1})), \quad c = \|\underline{v}_{j+1} - P_j(\underline{v}_{j+1})\|^{-1} \ ,$$

and by Thm. 8.3(ii) of Ch. 1 the right side here is $c(\underline{u}_i \cdot \underline{v}_{j+1} - P(u_i) \cdot \underline{v}_{j+1}) = c(\underline{u}_i \cdot \underline{v}_{j+1} - \underline{u}_i \cdot \underline{v}_{j+1}) = 0$, so $\underline{u}_{j+1}$ is orthogonal to $\underline{u}_1, \ldots, \underline{u}_j$, hence (since the inductive hypothesis guarantees that $\underline{u}_1, \ldots, \underline{u}_j$ are orthonormal) $\underline{u}_1, \ldots, \underline{u}_{j+1}$ are orthonormal and by 5.1 are l.i. vectors in the $(j+1)$-dimensional space span$\{\underline{v}_1, \ldots, \underline{v}_{j+1}\}$, and so are a basis for span$\{\underline{v}_1, \ldots, \underline{v}_{j+1}\}$. Thus, in particular, span$\{\underline{u}_1, \ldots, \underline{u}_{j+1}\}$ = span$\{\underline{v}_1, \ldots, \underline{v}_{j+1}\}$ and the proof of 5.3 is complete.

5.4 Remark: Notice that in view of Lem. 5.2 the inductive definition (1) in the above proof can be written

$$\underline{u}_{j+1} = \| \underline{v}_{j+1} - \sum_{i=1}^{j} \underline{v}_{j+1} \cdot \underline{u}_i \, \underline{u}_i \|^{-1} (\underline{v}_{j+1} - \sum_{i=1}^{j} \underline{v}_{j+1} \cdot \underline{u}_i \, \underline{u}_i) \, .$$

The procedure of getting orthonormal vectors in this way is known as *Gram-Schmidt orthogonalization*.

Finally, we want to introduce the important notion of orthogonal matrix.

5.5 Definition: An $n \times n$ matrix $Q = (q_{ij})$ is orthogonal if the columns $q_1, \ldots, q_n$ of Q are orthonormal.

Observe that by definition Q orthogonal $\Rightarrow Q^T Q = I$, because by definition of matrix multiplication the entry of $Q^T Q$ in the i-th row and j-th column is $q_i \cdot q_j$ which is δ_{ij} because $q_1, \ldots, q_n$ are orthonormal. Notice that the fact the $Q^T Q = I$ means that $\det(Q^T Q) = 1$, and hence by properties P4, P5 we see that $(\det Q)^2 = 1$, or in other words $\det Q = \pm 1$. In particular, $\det Q \neq 0$, so Q^{-1} exists by Thm. 3.4. Indeed, since $Q^T Q = I$ we see that $Q^{-1} = Q^T$ and hence we must automatically have $Q Q^T = I$. That is (since $Q = (Q^T)^T$) we have

$$Q \text{ orthogonal} \iff Q^T \text{ orthogonal} \, ,$$

so, in fact, Q orthogonal implies that the *rows* of Q are orthonormal.

SECTION 5 EXERCISES

5.1 If S is the two-dimensional subspace of $\mathbb{R}^4$ spanned by the vectors $(1, 1, 0, 0)^T$, $(0, 0, 1, 1)^T$, find an orthonormal basis for S and find the matrix of the orthogonal projection of $\mathbb{R}^4$ onto S.

5.2 If S is the subspace of $\mathbb{R}^4$ spanned by the vectors $(1, 0, 0, 1)^T$, $(1, 1, 0, 0)^T$, $(0, 0, 1, 1)^T$, find an orthonormal basis for S, and find the matrix of the orthogonal projection of $\mathbb{R}^4$ onto S.

6 MATRIX REPRESENTATIONS OF LINEAR TRANSFORMATIONS

Suppose V is a nontrivial subspace of $\mathbb{R}^n$ and $T : V \to V$ is linear. Thus, $T(\lambda \underline{x} + \mu \underline{y}) = \lambda T(\underline{x}) + \mu T(\underline{y}) \, \forall \underline{x}, \underline{y} \in V$ and for all $\lambda, \mu \in \mathbb{R}$. We showed in Sec. 6 of Ch. 1 that, in case $V = \mathbb{R}^n$, T is represented by the $n \times n$ matrix A with j-th column $T(\underline{e}_j)$ in the sense that $T(\underline{x}) \equiv A\underline{x}$ for all $\underline{x} \in \mathbb{R}^n$. A similar result holds in the present more general case when V is an arbitrary nontrivial subspace of $\mathbb{R}^n$.

To describe this result we first need to introduce a little terminology.

6.1 Terminology: Let $\underline{v}_1, \ldots, \underline{v}_k$ be a basis for V. Then each $\underline{x} \in V$ can be uniquely represented $\sum_{j=1}^{k} x_j \underline{v}_j$. The vector $\xi = (x_1, \ldots, x_k)^{\mathrm{T}} \in \mathbb{R}^k$ is referred to as *the coordinates of $\underline{x}$ relative to the given basis $\underline{v}_1, \ldots, \underline{v}_k$.*

Observe that $\underline{x} = \sum_{j=1}^{k} x_j \underline{v}_j \Rightarrow T(\underline{x}) = \sum_{j=1}^{n} x_j T(\underline{v}_j)$ by linearity of T, and since $T(\underline{v}_j) \in V$ and $\underline{v}_1, \ldots, \underline{v}_k$ is a basis for V we can express $T(\underline{v}_j)$ as a unique linear combination of $\underline{v}_1, \ldots, \underline{v}_k$; thus

6.2
$$T(\underline{v}_j) = \sum_{i=1}^{k} a_{ij} \underline{v}_i \text{ for some unique } a_{ij} \in \mathbb{R} .$$

The $k \times k$ matrix $A = (a_{ij})$ is called *the matrix of T relative to the given basis $\underline{v}_1, \ldots, \underline{v}_n$.*

Observe that, still with $\underline{x} = \sum_{j=1}^{k} x_j \underline{v}_j$, we then have $T(\underline{x}) = \sum_{j=1}^{k} x_j T(\underline{v}_j) = \sum_{j=1}^{k} \sum_{i=1}^{k} a_{ij} x_j \underline{v}_i = \sum_{i=1}^{k} (\sum_{j=1}^{k} a_{ij} x_j) \underline{v}_i$. Thus if $\underline{x}$ has coordinates $\xi = (x_1, \ldots, x_k)^{\mathrm{T}}$ relative to $\underline{v}_1, \ldots, \underline{v}_k$ then $T(\underline{x})$ has coordinates $A\xi$ relative to the same basis $\underline{v}_1, \ldots, \underline{v}_k$, where $A = (a_{ij})$ is the matrix of T relative to the given basis $\underline{v}_1, \ldots, \underline{v}_k$ as in 6.2.

7   EIGENVALUES AND THE SPECTRAL THEOREM

Let $A = (a_{ij})$ be an $n \times n$ (real) matrix. We begin with the definition of eigenvalue of A.

Observe that the expression $\det(A - \lambda I)$ is a degree n polynomial in the variable λ and the coefficient of the top degree term λ^n is $(-1)^n$. Thus, $\det(A - \lambda I) = (-1)^n(\lambda^n + \sum_{j=0}^{n-1} a_j \lambda^j)$ for some real numbers $a_1, \ldots, a_{n-1}$.

One can check this directly in case $n = 2$ (in which case $\det(A - \lambda I) = (a_{11} - \lambda)(a_{22} - \lambda) - a_{12}a_{21} = \lambda^2 - (a_{11} + a_{22})\lambda + (a_{11}a_{22} - a_{12}a_{21}))$ and it can easily be checked, in general, by induction on n.

Notice that then the fundamental theorem of algebra tells us that there are (possibly complex) numbers $\lambda_1, \ldots, \lambda_n$ such that

7.1
$$\det(A - \lambda I) \equiv (-1)^n(\lambda - \lambda_1) \cdots (\lambda - \lambda_n), \quad \forall \lambda .$$

7.2 Definition: The numbers $\lambda_1, \ldots, \lambda_n$ (which are the roots of the equation $\det(A - \lambda I) = 0$) are called eigenvalues of the matrix A.

There are two important properties of such eigenvalues.

7.3 Lemma. *Let $\lambda_1, \ldots, \lambda_n$ be the eigenvalues of A as in 7.2 above. Then:*

(i) $\det A = \lambda_1 \cdots \lambda_n$ *(i.e., $\det A =$ the product of the eigenvalues of A);*

(ii) *for any $\lambda \in \mathbb{R}$: λ an eigenvalue of A $\iff \exists \underline{v} \in \mathbb{R}^n \setminus \{\underline{0}\}$ with $A\underline{v} = \lambda \underline{v}$.*

7.4 Remark: Any $\underline{v} \in \mathbb{R}^n \setminus \{\underline{0}\}$ with $A\underline{v} = \lambda \underline{v}$ as in (ii) above is called *an eigenvector of A corresponding the eigenvalue* λ. Notice we only here consider eigenvectors corresponding to real eigenvalues λ; it would also be possible to consider eigenvectors corresponding to complex eigenvalues λ, although in that case the corresponding eigenvectors would be in $\mathbb{C}^n \setminus \{\underline{0}\}$ rather than in $\mathbb{R}^n \setminus \{\underline{0}\}$.

Proof of 7.3: First note that (i) is proved simply by setting $\lambda = 0$ in the identity 7.1.

To prove (ii), simply observe that, for any real λ, $\det(A - \lambda I) = 0 \iff N(A - \lambda I) \neq \{\underline{0}\} \iff \exists \underline{v} \in \mathbb{R}^n \setminus \{\underline{0}\}$ with $(A - \lambda I)\underline{v} = \underline{0}$ by Thm. 3.4 of the present chapter.

We can now state and prove the spectral theorem.

7.4 Theorem (Spectral Theorem.) *Let $A = (a_{ij})$ be a symmetric $n \times n$ matrix. Then there is an orthonormal basis $\underline{v}_1, \ldots, \underline{v}_n$ for $\mathbb{R}^n$ consisting of eigenvectors of A.*

Proof: The proof is by induction on n. It is trivial in the case $n = 1$ because in that case the matrix A is just a real number and $\mathbb{R}^n$ is the real line $\mathbb{R}$. So assume $n \geq 2$ and as an inductive hypothesis assume the theorem is true with $n - 1$ in place of n.

Let $\mathcal{A}(\underline{x})$ be the quadratic form given by the $n \times n$ matrix A. Thus,

$$\mathcal{A}(\underline{x}) = \sum_{i,j=1}^{n} a_{ij} x_i x_j \, ,$$

and observe that we can write

(1) $$\mathcal{A}(\underline{x}) = \|\underline{x}\|^2 \mathcal{A}(\widehat{x}), \quad \widehat{x} = \|\underline{x}\|^{-1} \underline{x} \in S^{n-1}, \quad \underline{x} \in \mathbb{R}^n \setminus \{\underline{0}\} \, .$$

By Thm. 2.6 of Ch. 2 we know that $\mathcal{A}|S^{n-1}$ attains its minimum value m at some point $\underline{v}_1 \in S^{n-1}$, so, by (1), for $\underline{x} \in \mathbb{R}^n \setminus \{\underline{0}\}$

(2) $$\|\underline{x}\|^{-2} \mathcal{A}(\underline{x}) = \mathcal{A}(\widehat{x}) \geq m = \|\underline{x}\|^{-2} \mathcal{A}(\underline{x})|_{\underline{x}=\underline{v}_1}$$

and hence the function $\|\underline{x}\|^{-2}\mathcal{A}(\underline{x}) : \mathbb{R}^n \setminus \{\underline{0}\} \to \mathbb{R}$ actually attains a minimum at the point $\underline{v}_1$ and hence (see the discussion following 7.1 of Ch. 2) must have zero gradient at this point. But direct computation shows that the i-th partial derivative of the function $\|\underline{x}\|^{-2}\mathcal{A}(\underline{x})$ is

$$2\|\underline{x}\|^{-2}\Big(\sum_{j=1}^{n} a_{ij} x_j - \|\underline{x}\|^{-3} \sum_{k,\ell=1}^{n} a_{k\ell} x_k x_\ell \, x_i\Big) \, ,$$

so at $\underline{x} = \underline{v}_1 = (v_{11}, v_{12}, \ldots, v_{1n})^{\mathrm{T}}$ we get

$$\sum_{j=1}^{n} a_{ij} v_{1j} - m v_{1i} = 0, \quad i = 1, \ldots, n ,$$

which in matrix terms says

$$A\underline{v}_1 = m\underline{v}_1 ,$$

so, by 7.3(ii), we have shown that $\underline{v}_1$ is an eigenvector corresponding to eigenvalue m.

Let T be the linear transformation determined by the matrix A, so that

$$T(\underline{x}) = A\underline{x}, \quad \underline{x} \in \mathbb{R}^n ,$$

and let $V = \mathrm{span}\{\underline{v}_1\}$. Then dim $V = 1$ and by Thm. 8.1(iii) we know that dim $V^\perp = n - 1$, and it is a straightforward exercise (Exercise 7.1(b) below) to check that $T : V^\perp \to V^\perp$. So let $\underline{w}_1, \ldots, \underline{w}_{n-1}$ be an orthonormal basis for $V^\perp$ and let $A_0 = (a_{ij}^{(0)})$ be the $(n-1) \times (n-1)$ matrix of $T|V^\perp$ relative to the basis $\underline{w}_1, \ldots, \underline{w}_{n-1}$ as discussed in Sec. 6 above. Observe that then by 6.2 we have

$$T(\underline{w}_j) = \sum_{k=1}^{n-1} a_{kj}^{(0)} \underline{w}_k, \quad j = 1, \ldots, n - 1 ,$$

and by taking the dot product of each side with $\underline{w}_i$ we get (using Exercise 7.1(a) below)

$$a_{ij}^{(0)} = \underline{w}_i \cdot T(\underline{w}_j) = \underline{w}_j \cdot T(\underline{w}_i) = a_{ji}^{(0)} ,$$

so that $a_{ij}^{(0)} = a_{ji}^{(0)}$. Thus, $(a_{ij}^{(0)})$ is an $(n-1) \times (n-1)$ symmetric matrix, and we can use the inductive hypothesis to find an orthonormal basis $\underline{u}_2, \ldots, \underline{u}_n$ of $\mathbb{R}^{n-1}$ and corresponding $\lambda_2, \ldots, \lambda_n$ such that

$$A_0 \underline{u}_j = \lambda_j \underline{u}_j, \quad j = 2, \ldots, n ,$$

so, with $\underline{u}_j = (u_{j1}, \ldots, u_{j\,n-1})^{\mathrm{T}}$,

(2)
$$\sum_{i=1}^{n-1} a_{ki}^{(0)} u_{ji} = \lambda_j u_{jk}, \quad j, k = 1, \ldots, n - 1 .$$

We then let $\underline{v}_j = \sum_{i=1}^{n-1} u_{ji} \underline{w}_i$ for $j = 2, \ldots, n$ and observe that by (2)

$$A\underline{v}_j = \sum_{i=1}^{n-1} u_{ji} A\underline{w}_i = \sum_{i=1}^{n-1} u_{ji} T(\underline{w}_i) = \sum_{i,k=1}^{n-1} u_{ji} a_{ki}^{(0)} \underline{w}_k = \lambda_j \sum_{k=1}^{n-1} u_{jk} \underline{w}_k = \lambda_j \underline{v}_j ,$$

for $j = 2, \ldots, n$, so $\underline{v}_j$ are eigenvectors of A for $j = 2, \ldots, n$. Finally, observe that since by construction $\underline{v}_2, \ldots, \underline{v}_n \in V^\perp$ and $\underline{v}_1 \in V$ we have that $\underline{v}_1, \ldots, \underline{v}_n$ are an orthonormal set of vectors in $\mathbb{R}^n$, and hence (since orthonormal vectors are automatically l.i. by Rem. 5.1 of the present Chapter) $\underline{v}_1, \ldots, \underline{v}_n$ are an orthonormal basis for $\mathbb{R}^n$.

SECTION 7 EXERCISES

7.1 Let $T : \mathbb{R}^n \to \mathbb{R}^n$ be a linear transformation with the property that the matrix of T (i.e., the matrix Q such that $T(\underline{x}) = Q\underline{x}$ for all $\underline{x} \in \mathbb{R}^n$) is symmetric. (i.e., $Q = (q_{ij})$ with $q_{ij} = q_{ji}$). Prove:

(a) $\underline{x} \cdot T(\underline{y}) = \underline{y} \cdot T(\underline{x})$ for all $\underline{x}, \underline{y} \in \mathbb{R}^n$.

(b) If $\underline{v}$ is a nonzero vector such that $T(\underline{v}) = \lambda \underline{v}$ for some $\lambda \in \mathbb{R}$, and if ℓ is the one-dimensional subspace spanned by $\underline{v}$, then $T(\ell^\perp) \subset \ell^\perp$.

7.2 Let A be an $n \times n$ symmetric matrix. Prove there is an $n \times n$ orthogonal matrix Q (i.e., $Q^T Q = I$) such that $Q^T A Q$ is a diagonal matrix.

Hint: The Spectral Theorem proved above says that there is an orthonormal basis $\underline{v}_1, \ldots, \underline{v}_n$ for $\mathbb{R}^n$ such that $A\underline{v}_j = \lambda_j \underline{v}_j$ for each j (where $\lambda_j \in \mathbb{R}$ for each j).

CHAPTER 4

More Analysis in $\mathbf{R}^n$

1   CONTRACTION MAPPING PRINCIPLE

First we define the notion of *contraction*. Given a mapping $f : E \to \mathbb{R}^n$, where $E \subset \mathbb{R}^n$, we say that f is a contraction if there is a constant $\theta \in (0, 1)$ such that

1.1
$$\|f(\underline{x}) - f(\underline{y})\| \le \theta \|\underline{x} - \underline{y}\|, \quad \forall \, \underline{x}, \underline{y} \in E \, .$$

Theorem (Contraction Mapping Theorem.) *Let E be a closed subset of $\mathbb{R}^n$ and f a contraction mapping of E into E. (Thus, $f(E) \subset E$ and f satisfies 1.1.) Then f has a fixed point, i.e., there is a $\underline{z} \in E$ such that $f(\underline{z}) = \underline{z}$.*

Remark: The proof we give below is *constructive*, i.e., not only will the proof show that there is a fixed point $\underline{z}$, but at the same time it actually gives a practical method for finding (or at least approximating arbitrarily closely) such a point $\underline{z}$.

Proof of the Contraction Mapping Theorem: We inductively construct a sequence $\{\underline{x}_k\}_{k=0,1,2,\dots}$ of points of E as follows:

(i) Take $\underline{x}_0 \in E$ arbitrary.

(ii) Assuming that $k \ge 1$ and that $\underline{x}_0, \dots, \underline{x}_{k-1}$ are already chosen, define

$$\underline{x}_k = f(\underline{x}_{k-1}) \, .$$

Observe that then we have, using 1.1,

$$\|\underline{x}_{k+1} - \underline{x}_k\| = \|f(\underline{x}_k) - f(\underline{x}_{k-1})\| \le \theta \|\underline{x}_k - \underline{x}_{k-1}\|, \quad k \ge 1 \, ,$$

and by mathematical induction we have

$$\|\underline{x}_{k+1} - \underline{x}_k\| \le \theta^k \|\underline{x}_1 - \underline{x}_0\|, \quad k \ge 1 \, ,$$

and so if $\ell > k \ge 1$ we have

$$\|\underline{x}_\ell - \underline{x}_k\| = \|\textstyle\sum_{j=k}^{\ell-1}(\underline{x}_{j+1} - \underline{x}_j)\| \le \sum_{j=k}^{\ell-1}\|\underline{x}_{j+1} - \underline{x}_j\|$$
$$\le (\textstyle\sum_{j=k}^{\ell-1}\theta^j)\|\underline{x}_1 - x_0\| = (\sum_{j=k}^{\ell-1}\theta^j)\|f(\underline{x}_0) - x_0\| \, .$$

On the other hand, $\sum_{j=k}^{\infty} \theta^j$ is a convergent geometric series with first term θ^k and common ratio θ, so it has sum $\theta^k/(1 - \theta)$, and hence the above inequality gives

(*)
$$\|\underline{x}_\ell - \underline{x}_k\| \le (1 - \theta)^{-1}\|f(\underline{x}_0) - x_0\|\theta^k, \quad \ell > k \ge 1 \, .$$

In particular, with $k = 1$ we have

$$\|\underline{x}_\ell\| = \|\underline{x}_\ell - \underline{x}_1 + \underline{x}_1\| \le \|\underline{x}_\ell - \underline{x}_1\| + \|\underline{x}_1\| \le (1 - \theta)^{-1} \|f(\underline{x}_0) - \underline{x}_0\| + \|f(\underline{x}_0)\|$$

for all $\ell > 1$, so the sequence $\{\underline{x}_k\}_{k=1,2,\dots}$ is bounded, and by the Bolzano-Weierstrass theorem there is a convergent subsequence $\{\underline{x}_{k_j}\}_{j=1,2,\dots}$. Since E is closed the limit $\lim_{j\to\infty} \underline{x}_{k_j}$ is a point of E. Notice that we can also take $\ell = k_j$ in (*), giving

(**) $$\|\underline{x}_{k_j} - \underline{x}_k\| \le (1 - \theta)^{-1} \|f(\underline{x}_0) - \underline{x}_0\| \theta^k, \quad j > k \ge 1\,,$$

and hence, letting $\underline{z} = \lim_{j\to\infty} \underline{x}_{k_j}$, and taking limits in (**), we see that

$$\|\underline{z} - \underline{x}_k\| \le (1 - \theta)^{-1} \|f(\underline{x}_0) - \underline{x}_0\| \theta^k, \quad \forall\, k \ge 1\,,$$

so that, in fact (since $\theta^k \to 0$ as $k \to \infty$), the whole sequence $\underline{x}_k$ (and not just the subsequence $\underline{x}_{k_j}$) converges to $\underline{z}$. Then by continuity of f at the point $\underline{z}$ we have $\lim f(\underline{x}_k) = f(\underline{z})$. But, on the other hand, $f(\underline{x}_k) = \underline{x}_{k+1}$ so we must also have $\lim f(\underline{x}_k) = \lim \underline{x}_{k+1} = \underline{z}$. Thus, $f(\underline{z}) = \underline{z}$.

SECTION 1 EXERCISES

1.1 Show that the system of two nonlinear equations

$$x^2 - y^2 + 4x - 1 = 0$$
$$x\left(\tfrac{1}{2}x^2 + y^2\right) + 4y + 1 = 0$$

has a solution in the disk $x^2 + y^2 \le 1$.

Hint: Define $f(x, y) = \left(x + \tfrac{1}{4}(x^2 - y^2) - \tfrac{1}{4},\, y + \tfrac{1}{4}(\tfrac{1}{2}x^2 + y^2)x + \tfrac{1}{4}\right)^{\mathrm{T}}$ and show that $F(x, y) = (x, y)^{\mathrm{T}} - f(x, y)$ has a fixed point in the disk.

2   INVERSE FUNCTION THEOREM

We here prove the inverse function theorem, which guarantees that C^1 function $f : U \to \mathbb{R}^n$ is locally C^1 invertible near points where the Jacobian matrix is nonsingular. More precisely:

Theorem (Inverse Function Theorem.) *Suppose $U \subset \mathbb{R}^n$ is open, $f : U \to \mathbb{R}^n$ is C^1, and $\underline{x}_0 \in U$ is such that $Df(\underline{x}_0)$ is nonsingular (i.e., $(Df(\underline{x}_0))^{-1}$ exists). Then there are open sets V, W containing $\underline{x}_0$ and $f(\underline{x}_0)$, respectively, such that $f|V$ is a 1:1 map of V onto W with C^1 inverse $g : W \to V$.*

Proof: We first consider the special case when $Df(\underline{x}_0) = I$. (The general case follows directly by applying this special case to the function $\tilde{f}(\underline{x}) \equiv (Df(\underline{x}_0))^{-1} f(\underline{x})$.)

Let $\varepsilon \in (0, \tfrac{1}{2}]$. Observe that since U is open and $Df(\underline{x})$ is continuous and $= I$ at $\underline{x} = \underline{x}_0$ we have a $\rho > 0$ such that $B_\rho(\underline{x}_0) \subset U$ and $\|Df(\underline{x}) - I\| < \varepsilon \le \tfrac{1}{2}$ for $\|\underline{x} - \underline{x}_0\| < \rho$, so, in particular, $\|Df(\underline{x})\underline{v}\| = \|\underline{v} + (Df(\underline{x}) - I)\underline{v}\| \ge \|\underline{v}\| - \|(Df(\underline{x}) - I)\underline{v}\| \ge \|\underline{v}\| - $

$\|(Df(\underline{x}) - I)\|\|\underline{v}\| \geq \|\underline{v}\| - \frac{1}{2}\|\underline{v}\| = \frac{1}{2}\|\underline{v}\|$ for any $\underline{v} \in \mathbb{R}^n$ and hence the null space $N(Df(\underline{x}))$ is trivial. Thus, the above choice of ρ ensures that $(Df(\underline{x}))^{-1}$ exists for each $\underline{x} \in B_\rho(\underline{x}_0)$.

Next observe that with the same $\rho > 0$ as in the above paragraph we have, by the fundamental theorem of calculus, $\forall \underline{x}, \underline{y} \in B_\rho(\underline{x}_0)$

$$(\underline{y} - f(\underline{y})) - (\underline{x} - f(\underline{x})) = \int_0^1 \tfrac{d}{dt} F(\underline{x} + t(\underline{y} - \underline{x}))\, dt, \quad F(\underline{x}) = \underline{x} - f(\underline{x}),$$

which by the Chain Rule gives

$$(\underline{y} - f(\underline{y})) - (\underline{x} - f(\underline{x})) = \int_0^1 (I - Df(\underline{x} + t(\underline{y} - \underline{x}))) \cdot (\underline{y} - \underline{x})\, dt,$$

and hence

(1)
$$\begin{aligned}
\|(\underline{y} - f(\underline{y})) - (\underline{x} - f(\underline{x}))\| &= \| \int_0^1 (I - Df(\underline{x} + t(\underline{y} - \underline{x}))) \cdot (\underline{y} - \underline{x})\, dt\| \\
&\leq \int_0^1 \|(I - Df(\underline{x} + t(\underline{y} - \underline{x}))) \cdot (\underline{y} - \underline{x})\|\, dt \\
&\leq \int_0^1 \|I - Df(\underline{x} + t(\underline{y} - \underline{x}))\|\, \|\underline{y} - \underline{x}\|\, dt \\
&\leq \varepsilon\|\underline{y} - x\| \quad \forall \underline{x}, \underline{y} \in B_\rho(\underline{x}_0),
\end{aligned}$$

and hence

(2)
$$(1 - \varepsilon)\|\underline{x} - \underline{y}\| \leq \|f(\underline{x}) - f(\underline{y})\| \leq (1 + \varepsilon)\|\underline{x} - \underline{y}\|, \quad \forall \underline{x}, \underline{y} \in B_\rho(\underline{x}_0).$$

We are going to prove the theorem with $V = B_\rho(\underline{x}_0)$. We first claim that if $\underline{z}$ is an arbitrary point of the open ball $B_\rho(\underline{x}_0)$ and if $\sigma \in (0, \rho - \|\underline{z} - \underline{x}_0\|)$ then

(3)
$$B_{\sigma/2}(f(\underline{z})) \subset f(B_\sigma(\underline{z})) \subset f(B_\rho(\underline{x}_0)).$$

Notice that the second inclusion here is trivial because of the inclusion $\overline{B}_\sigma(\underline{z}) \subset B_\rho(\underline{x}_0)$ which we check as follows: $\underline{x} \in \overline{B}_\sigma(\underline{z}) \Rightarrow \|\underline{x} - \underline{x}_0\| = \|\underline{x} - \underline{z} + \underline{z} - \underline{x}_0\| \leq \|\underline{x} - \underline{z}\| + \|\underline{z} - \underline{x}_0\| \leq \sigma + \|\underline{z} - \underline{x}_0\| < \rho$ because $\sigma < \rho - \|\underline{z} - \underline{x}_0\|$. Thus, we have only to prove the first inclusion in (3). To prove this we have to show that for each vector $\underline{v} \in B_{\sigma/2}(f(\underline{z}))$ there is an $\underline{x}_{\underline{v}} \in B_\sigma(\underline{z})$ with $f(\underline{x}_{\underline{v}}) = \underline{v}$. We check this using the contraction mapping principle on the closed ball $\overline{B}_\sigma(\underline{z})$: We define $F(\underline{x}) = \underline{x} - f(\underline{x}) + \underline{v}$, and note that

$$\begin{aligned}
\|F(\underline{x}) - \underline{z}\| &= \|(\underline{x} - f(\underline{x})) - (\underline{z} - f(\underline{z})) + \underline{v} - f(\underline{z})\| \\
&\leq \|(\underline{x} - f(\underline{x})) - (\underline{z} - f(\underline{z}))\| + \|\underline{v} - f(\underline{z})\| < \tfrac{1}{2}\sigma + \tfrac{1}{2}\sigma = \sigma,
\end{aligned}$$

for all $\underline{x} \in \overline{B}_\sigma(\underline{z})$, so F maps the closed ball $\overline{B}_\sigma(\underline{z})$ into the corresponding open ball $B_\sigma(\underline{z})$ and since $F(x) - F(\underline{y}) = (\underline{x} - f(\underline{x})) - (\underline{y} - f(\underline{y}))$ we see by (1) that F is a contraction mapping of $\overline{B}_\sigma(\underline{z})$ into itself and so has a fixed point $\underline{x} \in \overline{B}_\sigma(\underline{z})$ by the contraction mapping principle. This fixed point $\underline{x}$ actually lies in the open ball $B_\sigma(\underline{z})$ because F maps $\overline{B}_\sigma(\underline{z})$ into $B_\sigma(\underline{z})$; thus $\underline{x} - f(\underline{x}) + \underline{v} = \underline{x}$, or in other words $f(\underline{x}) = \underline{v}$, with $\underline{x} \in B_\sigma(\underline{z})$, so $\underline{x}$ is the required point $\underline{x}_{\underline{v}} \in B_\sigma(\underline{z})$, and indeed (3) is proved. Notice that (3) in particular guarantees that the image $f(B_\rho(\underline{x}_0))$ is an open set, which we call W. By (2), f is a 1:1 map of $B_\rho(\underline{x}_0)$ onto this open set W. Let $g : W \to B_\rho(\underline{x}_0)$ be the inverse function and write $\underline{x} = g(\underline{u})$, $\underline{y} = g(\underline{v})$ ($\underline{u}$, $\underline{v} \in W$); then the first inequality in (2) can be written

(2)′ $$\|g(\underline{u}) - g(\underline{v})\| \le (1 - \varepsilon)^{-1}\|\underline{u} - \underline{v}\| \le 2\|\underline{u} - \underline{v}\|, \quad \underline{u}, \underline{v} \in W,$$

which guarantees that the inverse function g is continuous on W. To prove the inverse is differentiable at a given $\underline{v} \in W$, pick $\underline{y} \in B_\rho(\underline{x}_0)$ such that $f(\underline{y}) = \underline{v}$, let $A = Df(\underline{y})$. We claim that g is differentiable at $\underline{v}$ with $Dg(\underline{v}) = (Df(y))^{-1}$. To see this let $A = Df(y)$, and for given $\varepsilon \in (0, \frac{1}{4})$ pick $\delta > 0$ such that $B_\delta(\underline{v}) \subset W$ (which we can do since W is open), $B_\delta(\underline{y}) \subset B_\rho(\underline{x}_0)$, and

(4) $$0 < \|\underline{x} - \underline{y}\| < \delta \Rightarrow \|f(\underline{x}) - f(\underline{y}) - A(\underline{x} - \underline{y})\| < \varepsilon\|\underline{x} - \underline{y}\|,$$

which we can do since f is differentiable at $\underline{y}$. Recall that we already checked that $A^{-1}(= (Df(\underline{y}))^{-1})$ exists for all $\underline{y} \in B_\rho(\underline{x}_0)$, and then

$$\|g(\underline{u}) - g(\underline{v}) - A^{-1}(\underline{u} - \underline{v})\| = \|A^{-1}(A(g(\underline{u}) - g(\underline{v})) - (\underline{u} - \underline{v}))\|$$
$$= \|A^{-1}(A(x - y) - (f(\underline{x}) - f(\underline{y})))\|$$
(5) $$\le \|A^{-1}\|\,\|f(\underline{x}) - f(\underline{y}) - A(x - y)\|$$

for every $\underline{u} \in W$, where $\underline{x} = g(\underline{u})$ (i.e., $\underline{u} = f(\underline{x})$). But, since $\|\underline{x} - \underline{y}\| = \|g(\underline{u}) - g(\underline{v})\| \le 2\|\underline{u} - \underline{v}\|$ (by (2)′), we then have $\|\underline{u} - \underline{v}\| < \delta/2 \Rightarrow \|\underline{x} - \underline{y}\| < \delta$, which by (4) implies the expression on the right of (5) is $\le \varepsilon\|A^{-1}\|\,\|\underline{x} - \underline{y}\| = \varepsilon\|A^{-1}\|\,\|g(\underline{u}) - g(\underline{v})\| \le 2\varepsilon\|A^{-1}\|\,\|u - v\|$. That is, we have proved

$$\|\underline{u} - \underline{v}\| < \delta/2 \Rightarrow \|g(\underline{u}) - g(\underline{v}) - A^{-1}(\underline{u} - \underline{v})\| < 2\varepsilon\|A^{-1}\|\,\|\underline{u} - \underline{v}\|,$$

which proves that g is differentiable at $\underline{v}$ with $Dg(\underline{v}) = A^{-1}$, as claimed.

Finally, we have to check the *continuity* of the Jacobian matrix $Dg(\underline{u})$ for $\underline{u} \in W$; but $f(g(\underline{u}))) \equiv \underline{u}$ for $\underline{u} \in W$, so by the chain rule we have $Df(g(\underline{u}))Dg(\underline{u}) = I$ for every $\underline{u} \in W$; i.e., $Dg(\underline{u}) = (Df(g(\underline{u})))^{-1}$, which is a continuous function of $\underline{u}$ because g is a continuous function of $\underline{u} \in W$ and Df is continuous on $B_\rho(\underline{x}_0)$, which in turn implies that $(Df)^{-1}$ is continuous on $B_\rho(\underline{x}_0)$. This completes the proof of the inverse function theorem.

3   IMPLICIT FUNCTION THEOREM

The implicit function theorem considers C^1 mappings from an open set U in $\mathbb{R}^n$ into $\mathbb{R}^m$, with $m < n$. It is convenient notationally to think of $\mathbb{R}^n$ as the same as the product $\mathbb{R}^{n-m} \times \mathbb{R}^m$, and accordingly, points in $\mathbb{R}^n$ should be written $\begin{pmatrix} \underline{x} \\ \underline{y} \end{pmatrix}$ where $\underline{x} \in \mathbb{R}^{n-m}$ and $\underline{y} \in \mathbb{R}^m$; this is somewhat cumbersome so we instead abbreviate it by writing $[\underline{x}, \underline{y}]$ (with square brackets). Likewise, a function f on $\mathbb{R}^n$ should be written $f\begin{pmatrix} \underline{x} \\ \underline{y} \end{pmatrix}$, again too cumbersome, so we'll write $f[\underline{x}, \underline{y}]$. After a while you'll get comfortable with this, and start fudging notation by simply writing $f(\underline{x}, \underline{y})$ (with round brackets) instead of $f[\underline{x}, \underline{y}]$. Note that if $f[\underline{x}, \underline{y}]$ is a C^1 function then we use the notation $D_{\underline{x}} f[\underline{x}, \underline{y}] = (\frac{\partial f[\underline{x}, \underline{y}]}{\partial x_1}, \ldots, \frac{\partial f[\underline{x}, \underline{y}]}{\partial x_{n-m}})$ and $D_{\underline{y}} f[\underline{x}, \underline{y}] = (\frac{\partial f[\underline{x}, \underline{y}]}{\partial y_1}, \ldots, \frac{\partial f[\underline{x}, \underline{y}]}{\partial y_m})$, and so in this notation the Jacobian matrix of f is $(D_{\underline{x}} f[\underline{x}, \underline{y}], D_{\underline{y}} f[\underline{x}, \underline{y}])$.

Implicit Function Theorem. *Let U be an open subset of $\mathbb{R}^n$, $G : U \to \mathbb{R}^m$ a C^1 mapping with $m < n$, let $S = \{[\underline{x}, \underline{y}] \in U : G[\underline{x}, \underline{y}] = \underline{0}\}$, and let $[\underline{a}, \underline{b}]$ be a point in S such that the matrix $D_{\underline{y}} G[\underline{x}, \underline{y}]$ (that's the $m \times m$ matrix with j-th column $\frac{\partial G[\underline{x}, \underline{y}]}{\partial y_j}$) is nonsingular at the point $[\underline{a}, \underline{b}]$. Then there are open sets $V \subset \mathbb{R}^n$ and $W \subset \mathbb{R}^{n-m}$ with $[\underline{a}, \underline{b}] \in V$ and $\underline{a} \in W$ and a C^1 map $h : W \to \mathbb{R}^m$ such that $S \cap V = \text{graph } h$.*

Note: As usual, graph $h = \{[\underline{x}, \underline{y}] : \underline{x} \in W \text{ and } \underline{y} = h(\underline{x})\}$, that is, $\{[\underline{x}, h(\underline{x})] : \underline{x} \in W\}$.

Proof of the Implicit Function Theorem: The proof is rather easy: it is simply a clever application of the inverse function theorem. We start by defining a map $f : U \to \mathbb{R}^n$ by simply taking $f[\underline{x}, \underline{y}] = [\underline{x}, G[\underline{x}, \underline{y}]]$. Of course this is a C^1 map because G is C^1. Note that the mapping f leaves the first $n - m$ coordinates (i.e., the $\underline{x}$ coordinates) of each point $[\underline{x}, \underline{y}]$ fixed. Also, by direct calculation the Jacobian matrix $Df[\underline{a}, \underline{b}]$ (which is $n \times n$) has the block form

$$\begin{pmatrix} I_{(n-m) \times (n-m)} & O \\ D_{\underline{x}} G[\underline{a}, \underline{b}] & D_{\underline{y}} G[\underline{a}, \underline{b}] \end{pmatrix}$$

and hence det $Df[\underline{a}, \underline{b}] = \det D_{\underline{y}} G[\underline{a}, \underline{b}] \neq 0$ so the inverse function theorem implies that there are open sets V, Q in $\mathbb{R}^n$ with $[\underline{a}, \underline{b}] \in V$, $f[\underline{a}, \underline{b}] = [\underline{a}, G[\underline{a}, \underline{b}]] = [\underline{a}, \underline{0}] \in Q$ such that $f : V \to Q$ is 1:1 and onto and the inverse $g : Q \to V$ is C^1. Since $f[\underline{x}, \underline{y}] \equiv [\underline{x}, G[\underline{x}, \underline{y}]]$ (i.e., f fixes the first $n - m$ components $\underline{x}$ of the point $[\underline{x}, \underline{y}] \in \mathbb{R}^n$), g must have the form

(1) $\qquad\qquad g[\underline{x}, \underline{y}] = [\underline{x}, H[\underline{x}, \underline{y}]], \quad$ with H a C^1 function on Q.

Now we define $W = \{\underline{x} \in \mathbb{R}^{n-m} : [\underline{x}, \underline{0}] \in Q \cap (\mathbb{R}^{n-m} \times \{\underline{0}\})\}$ (it is easy to check that W is then open in $\mathbb{R}^{n-m}$ because Q is open in $\mathbb{R}^n$), and let $h(\underline{x}) = H[\underline{x}, \underline{0}]$ with H as in (1), so that

$$g[\underline{x}, \underline{0}] = [\underline{x}, h(\underline{x})], \quad \underline{x} \in W, \text{ where } h(\underline{x}) = H[\underline{x}, \underline{0}] \text{ with } H \text{ as in (1)},$$

and then observe that $[\underline{x}, \underline{0}] = f(g[\underline{x}, \underline{0}]) = f[\underline{x}, h(\underline{x})] = [\underline{x}, G[\underline{x}, h(\underline{x})]]$; that is, $G[\underline{x}, h(\underline{x})] = \underline{0}$, so we've shown that graph $h \subset S \cap V$. To check equality, we just have to check the reverse inclusion $S \cap V \subset \text{graph } h$ as follows:

$$[\underline{x}, \underline{z}] \in S \cap V \Rightarrow G[\underline{x}, \underline{z}] = \underline{0} \Rightarrow f[\underline{x}, \underline{z}] = [\underline{x}, \underline{0}] \Rightarrow$$
$$[\underline{x}, \underline{z}] = g(f[\underline{x}, \underline{z}]) = g[\underline{x}, \underline{0}] = [\underline{x}, h(\underline{x})],$$

so $h(\underline{x}) = \underline{z}$ and $[\underline{x}, \underline{z}] \in \text{graph } h$.

Finally, we give an alternative version of the implicit function theorem which we stated in Thm. 9.13 of Ch. 2 and which was used in the proof of the Lagrange multiplier Thm. 9.14 of Ch. 2.

Corollary. *If $k \in \{1, \ldots, n-1\}$, if $U \subset \mathbb{R}^n$ is open, if $g_1, \ldots, g_{n-k} : U \to \mathbb{R}$ are C^1, if $S = \{\underline{x} \in U : g_j(\underline{x}) = 0$ for each $j = 1, \ldots, n-k\}$ and if $M \neq \emptyset$, where*

$$M = \{\underline{y} \in S : \nabla_{\mathbb{R}^n} g_1(\underline{y}), \ldots, \nabla_{\mathbb{R}^n} g_{n-k}(\underline{y}) \text{ are l.i.}\},$$

then M is a k-dimensional C^1 manifold.

Proof: Let $m = n - k$, let $\underline{a} \in M$ and let $G = (g_1, \ldots, g_m)^{\mathrm{T}} : U \to \mathbb{R}^m$. The m rows of the $m \times n$ matrix $DG(\underline{a})$ are given to be l.i., so $DG(\underline{a})$ has rank m and we must we able find m l.i. columns $D_{j_1} G(\underline{a}), \ldots, D_{j_m} G(\underline{a})$. Reordering the coordinates if necessary, we can suppose without loss of generality that $j_1 = n - m + 1$, $j_2 = n - m + 2, \ldots, j_m = n$ (i.e., $j_1, \ldots, j_m$ are the last m integers between 1 and n), and we can relabel the coordinates of points of $\mathbb{R}^n$ as $[\underline{x}, \underline{y}]$ with $\underline{x} \in \mathbb{R}^{n-m}$, $\underline{y} \in \mathbb{R}^m$ and the point $\underline{a}$ can be written also as $[\underline{a}_0, \underline{b}_0]$ with $\underline{a}_0 \in \mathbb{R}^{n-m}$ and $\underline{b}_0 \in \mathbb{R}^m$. Then $D_{\underline{y}} G[\underline{x}, \underline{y}]$ is an $m \times m$ nonsingular matrix at the point $[\underline{a}_0, \underline{b}_0]$, so the Implicit Function Theorem applies with $[\underline{a}_0, \underline{b}_0]$ in place of $[\underline{a}, \underline{b}]$ and hence the Def. 9.2 of Ch. 2 is checked.

APPENDIX A

Introductory Lectures on Real Analysis

LECTURE 1: THE REAL NUMBERS

We assume without proof the usual properties of the integers: For example, that the integers are closed under addition and subtraction, that the principle of mathematical induction holds for the positive integers, and that 1 is the least positive integer.

We also assume the usual field and order properties for the real numbers $\mathbb{R}$. Thus, we accept without proof that the reals satisfy the *field axioms*, as follows:

F1 $a + b = b + a, ab = ba$ (commutativity)

F2 $a + (b + c) = (a + b) + c, a(bc) = (ab)c$ (associativity)

F3 $a(b + c) = ab + ac$ (distributive law)

F4 $\exists$ elements 0, 1 with $0 \neq 1$ such that $0 + a = a$ and $1 \cdot a = a$ $\forall a \in \mathbb{R}$ (additive and multiplicative identities)

F5 $\forall a \in \mathbb{R}, \exists$ an element $-a$ such that $a + (-a) = 0$ (additive inverse)

F6 $\forall a \in \mathbb{R}$ with $a \neq 0, \exists$ an element a^{-1} such that $aa^{-1} = 1$ (multiplicative inverse).

Terminology: $a + (-b), ab^{-1}$ are usually written $a - b$, $\frac{a}{b}$, respectively. Notice that the latter makes sense only for $b \neq 0$.

Notice that all the other standard algebraic properties of the reals follow from these. (See Exercise 1.1 below.)

We here also accept without proof that the reals satisfy the following *order axioms:*

01 For each $a \in \mathbb{R}$, *exactly one* of the possibilities $a > 0, a = 0, -a > 0$ holds.

02 $a > 0$ and $b > 0 \Rightarrow ab > 0$ and $a + b > 0$.

Terminology: $a > b$ means $a - b > 0$, $a \geq b$ means that either $a > b$ or $a = b$, $a < b$ means $b > a$, and $a \leq b$ means $b \geq a$.

We claim that all the other standard properties of inequalities follow from these and from F1–F6. (See Problem 1.2 below.)

Notice also that the above properties (i.e., F1–F6, O1, O2) all hold with the *rational numbers* $Q \equiv \{\frac{p}{q} : p, q$ are integers, $q \neq 0\}$ in place of $\mathbb{R}$. F1–F6 also hold with the complex numbers $\mathbb{C} = \{x + iy : x, y \in \mathbb{R}\}$ in place of $\mathbb{R}$, but inequalities like $a > b$ make no sense for complex numbers.

In addition to F1–F6, O1, O2 there is one further key property of the real numbers. To discuss it we need first to introduce some terminology.

Terminology: If $S \subset \mathbb{R}$ we say:

(1) S is *bounded above* if $\exists$ a number $K \in \mathbb{R}$ such that $x \leq K$ $\forall x \in S$. (Any such number K is called an *upper bound* for S.)

(2) S is *bounded below* if $\exists$ a number $k \in \mathbb{R}$ such that $x \geq k$ $\forall x \in S$. (Any such number k is called a *lower bound* for S.)

(3) S is *bounded* if it is *both* bounded above *and* bounded below. (This is equivalent to the fact that $\exists$ a positive real number L such that $|x| \leq L$ $\forall x \in S$.)

We can now introduce the final key property of the real numbers.

C. ("Completeness property of the reals"): If $S \subset \mathbb{R}$ is nonempty and bounded above, then S has a *least* upper bound.

Notice that the terminology "least upper bound" used here means exactly what it says: a number α is a least upper bound for a set $S \subset \mathbb{R}$ if

(i) $x \leq \alpha$ $\forall$ $x \in S$ i.e., α is *an* upper bound), and

(ii)   if β is any other upper bound for S, then $\alpha \leq \beta$ i.e., α is $\leq$ any other upper bound for S).

Such a least upper bound is *unique*, because if α_1, α_2 are both least upper bounds for S, the property (ii) implies that both $\alpha_1 \leq \alpha_2$ *and* $\alpha_2 \leq \alpha_1$, so $\alpha_2 = \alpha_1$. It therefore makes sense to speak of *the* least upper bound of S (also known as "the supremum" of S). The least upper bound of S will henceforth be denoted sup S. Notice that property C guarantees sup S *exists* if S is nonempty and bounded above.

Remark: If S is nonempty and bounded below, then it has a greatest lower bound (or "infimum"), which we denote inf S. One can *prove* the existence of inf S (if S is bounded below and nonempty) by using property C on the set $-S = \{-x : x \in S\}$. (See Exercise 1.5 below.)

We should be careful to distinguish between the maximum element of a set S (if it exists) and the supremum of S. Recall that we say a number α is the maximum of S (denoted max S) if (i)′ $x \le \alpha \; \forall \, x \in S$ (i.e., α is an upper bound for S), and

(ii)′ $\alpha \in S$.

These two properties say exactly that α is a upper bound for S and also one of the elements of S. Thus, clearly a maximum α of S, if it exists, satisfies both (i), (ii) and hence must agree with sup S. But keep in mind that max S may not exist, even if the set S is nonempty and bounded above: for example, if $S = (0, 1)$ ($= \{x \in \mathbb{R} : 0 < x < 1\}$), then sup $S = 1$, but max S does not exist, because $1 \notin S$.

Notice that of course any *finite* nonempty set $S \subset \mathbb{R}$ has a maximum. One can formally prove this by induction on the number of elements of the set S. (See Exercise 1.3 below.)

Using the above-mentioned properties of the integers and the reals it is now possible to give formal rigorous proofs of all the other properties of the reals which are used, even the ones which seem self-evident. For example, one can actually prove formally, using only the above properties, the fact that the set of positive integers are not bounded above. (Otherwise there would be a least upper bound α so, in particular, we would have $n \le \alpha$ for each positive integer n and hence, in particular, $n \le \alpha$ hence $n + 1 \le \alpha$ for each n, or in other words $n \le \alpha - 1$ for each positive integer n, contradicting the fact that α is the least upper bound!) Thus, we have shown rigorously, using only the axioms F1–F6, O1, O2, and C, that the positive integers are not bounded above. Thus,

$$\forall \text{ positive } a \in \mathbb{R}, \exists \text{ a positive integer } n \text{ with } n > a \quad \left(\text{i.e., } \frac{1}{n} < \frac{1}{a}\right).$$

This is referred to as "The Archimedean Property" of the reals.

Similarly, using only the axioms F1–F6, O1,2, and C, we can give a formal proof of all the basic properties of the real numbers—for example, in Problem 1.7 below you are asked to prove that square roots of positive numbers do indeed exist.

Final notes on the Reals: (1) We have *assumed* without proof all properties F1–F6, O1,O2 and C. In a more advanced course we could, starting only with the positive integers, give a rigorous *construction* of the real numbers, and *prove* all the properties F1–F6, O1, O2, and C. Furthermore, one can prove (in a sense that can be made precise) that the real number system is the *unique* field with all the above properties.

(2) You can of course freely use all the standard rules for algebraic manipulation of equations and inequalities involving the reals; normally you do not need to justify such things in terms of the axioms F1–F6, O1, O2 unless you are specifically asked to do so.

LECTURE 1 PROBLEMS

1.1 Using *only* properties F1–F6, prove

(i) $a \cdot 0 = 0 \; \forall \, a \in \mathbb{R}$

(ii) $ab = 0 \Rightarrow$ either $a = 0$ or $b = 0$

(iii) $\frac{a}{b} + \frac{c}{d} = \frac{ad+bc}{bd}$ $\forall a, b, c, d \in \mathbb{R}$ with $b \neq 0, d \neq 0$.

Note: In each case present your argument in steps, stating which of F1–F6 is used at each step.

Hint for (iii): First show the "cancellation law" that $\frac{x}{y} = \frac{xz}{yz}$ for any $x, y, z \in \mathbb{R}$ with $y \neq 0, z \neq 0$.

1.2 Using only properties F1–F6 and O1,O2, (and the agreed terminology) prove the following:

(i) $a > 0 \Rightarrow 0 > -a$ i.e., $-a < 0$)

(ii) $a > 0 \Rightarrow \frac{1}{a} > 0$

(iii) $a > b > 0 \Rightarrow \frac{1}{a} < \frac{1}{b}$

(iv) $a > b$ and $c > 0 \Rightarrow ac > bc$.

1.3 If S is a finite nonempty subset of $\mathbb{R}$, prove that max S exists. (Hint: Let n be the number of elements of S and try using induction on n.)

1.4 Given any number $x \in \mathbb{R}$, prove there is an integer n such that $n \leq x < n + 1$.

Hint: Start by proving there is a least integer $> x$.

Note: 1.4 Establishes rigorously the fact that every real number x can be written in the form $x =$ integer plus a remainder, with the remainder $\in [0, 1)$. The integer is often referred to as the "integer part of x." We emphasize once again that such properties are completely standard properties of the real numbers and can normally be used without comment; the point of the present exercise is to show that it is indeed possible to rigorously prove such standard properties by using the basic properties of the integers and the axioms F1–F6, O1,O2, C.

1.5 Given a set $S \subset \mathbb{R}$, $-S$ denotes $\{-x : x \in S\}$. Prove:

(i) S is bounded below if an only if $-S$ is bounded above.

(ii) If S is nonempty and bounded below, then inf S exists and $= -\sup(-S)$.

(Hint: Show that $\alpha = -\sup(-S)$ has the necessary 2 properties which establish it to be the greatest lower bound of S.)

1.6 If $S \subset \mathbb{R}$ is nonempty and bounded above, prove $\exists$ a sequence $a_1, a_2, \ldots$ of points in S with $\lim a_n = \sup S$.

Hint: In case sup $S \notin S$, let $\alpha = \sup S$, and for each integer $j \geq 1$ prove there is at least one element $a_j \in S$ with $\alpha > a_j > \alpha - \frac{1}{j}$.

1.7 Prove that every positive real number has a positive square root. (That is, for any $a > 0$, prove there is a real number $\alpha > 0$ such that $\alpha^2 = a$.)

Hint: Begin by observing that $S = \{x \in \mathbb{R} : x > 0 \text{ and } x^2 < a\}$ is nonempty and bounded above, and then argue that sup S is the required square root.

LECTURE 2: SEQUENCES OF REAL NUMBERS AND THE BOLZANO-WEIERSTRASS THEOREM

Let $a_1, a_2, \ldots$ be a sequence of real numbers; a_n is called the *n-th term* of the sequence. We sometimes use the abbreviation $\{a_n\}$ or $\{a_n\}_{n=1,2,\ldots}$

Technically, we should distinguish between *the sequence* $\{a_n\}$ and *the set of terms of the sequence*—i.e., the set $S = \{a_1, a_2 \ldots\}$. These are *not* the same: e.g., the sequence $1, 1, \ldots$ has infinitely many terms each equal to 1, whereas the set S is just the set $\{1\}$ containing one element.

Formally, a sequence is a *mapping* from the positive integers to the real numbers; the nth term a_n of the sequence is just the value of this mapping at the integer n. From this point of view—i.e., thinking of a sequence as a mapping from the integers to the real numbers—a sequence has a graph consisting of discrete points in $\mathbb{R}^2$, one point of the *graph* on each of the vertical lines $x = n$. Thus, for example, the sequence $1, 1, \ldots$ (each term $= 1$) has graph consisting of the discrete points marked "$\otimes$" in the following figure:

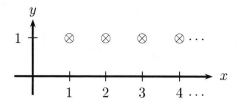

Figure A.1: Graph of the trivial sequence $\{a_n\}_{n=1,\ldots}$ where $a_n = 1 \; \forall n$.

Terminology: Recall the following terminology. A sequence $a_1, a_2, \ldots$ is:

(i)   *bounded above* if $\exists$ a real number K such that $a_n \leq K \; \forall$ integer $n \geq 1$.

(ii) *bounded below* if $\exists$ a real number k such that $a_n \geq k \; \forall$ integer $n \geq 1$.

(iii) *bounded* if it is *both* bounded above *and* bounded below. (This is equivalent to the fact that $\exists$ a real number L such that $|a_n| \leq L \; \forall$ integer $n \geq 1$.)

(iv) *increasing* if $a_{n+1} \geq a_n \; \forall$ integer $n \geq 1$.

(v)   *strictly increasing* if $a_{n+1} > a_n \; \forall$ integer $n \geq 1$.

(vi) *decreasing* if $a_{n+1} \leq a_n \; \forall$ integer $n \geq 1$.

(vii) *strictly decreasing* if $a_{n+1} < a_n \; \forall$ integer $n \geq 1$.

(viii) *monotone* if *either* the sequence is increasing or the sequence is decreasing.

(ix) We say the sequence *has limit* ℓ (ℓ a given real number) if for each $\varepsilon > 0$ there is an integer $N \geq 1$ such that

(∗) $|a_n - \ell| < \varepsilon \; \forall$ integer $n \geq N$.[3]

(x)   In case the sequence $\{a_n\}$ has limit ℓ we write

$$\lim a_n = \ell \text{ or } \lim_{n \to \infty} a_n = \ell \text{ or } a_n \to \ell \, .$$

(xi) We say the sequence $\{a_n\}$ *converges* (or "*is convergent*") if it has limit ℓ for some $\ell \in \mathbb{R}$.

Theorem 2.1. *If $\{a_n\}$ is monotone and bounded, then it is convergent. In fact, if $S = \{a_1, a_2 \ldots\}$ is the set of terms of the sequence, we have the following:*

(i)   *if $\{a_n\}$ is increasing and bounded then $\lim a_n = \sup S$.*

(ii) *if $\{a_n\}$ is decreasing and bounded then $\lim a_n = \inf S$.*

Proof: See Exercise 2.2. (Exercise 2.2 proves part (i), but the proof of part (ii) is almost identical.)

Theorem 2.2. *If $\{a_n\}$ is convergent, then it is bounded.*

Proof: Let $l = \lim a_n$. Using the definition (ix) above with $\varepsilon = 1$, we see that there exists an integer $N \geq 1$ such that $|a_n - l| < 1$ whenever $n \geq N$. Thus, using the triangle inequality, we have $|a_n| \equiv |(a_n - l) + l| \leq |a_n - l| + |l| < 1 + |l| \; \forall$ integer $n \geq N$. Thus,

$$|a_n| \leq \max\{|a_1|, \ldots |a_N|, |l| + 1\} \; \forall \text{ integer } n \geq 1 \, .$$

Theorem 2.3. *If $\{a_n\}$, $\{b_n\}$ are convergent sequences, then the sequences $\{a_n + b_n\}$, $\{a_n b_n\}$ are also convergent and*

(i) $\lim(a_n + b_n) = \lim a_n + \lim b_n$

(ii) $\lim(a_n b_n) = (\lim a_n) \cdot (\lim b_n) \, .$

In addition, if $b_n \neq 0$ and $\lim b_n \neq 0$, then

(iii) $$\lim \frac{a_n}{b_n} = \frac{\lim a_n}{\lim b_n} \, .$$

Proof: We prove (ii); the remaining parts are left as an exercise. First, since $\{a_n\}$, $\{b_n\}$ are convergent, the previous theorem tells us that there are constants $L, M > 0$ such that

(∗) $|a_n| \leq L$ and $|b_n| \leq M \; \forall$ positive integer n .

---

[3] Notice that (∗) is equivalent to $\ell - \varepsilon < a_n < \ell + \varepsilon \; \forall$ integers $n \geq N$.

Now let $l = \lim a_n$, $m = \lim b_n$ and note that by the triangle inequality and $(*)$,

$$
\begin{aligned}
|a_n b_n - lm| &\equiv |a_n b_n - l b_n + l b_n - lm| \\
&\equiv |(a_n - l)b_n + l(b_n - m)| \\
&\leq |a_n - l||b_n| + |l||b_n - m| \\
&\leq M|a_n - l| + |l||b_n - m| \;\forall n \geq 1 .
\end{aligned}
$$

$(**)$

On the other hand, for any given $\varepsilon > 0$ we use the definition of convergence, i.e., (ix) above) to deduce that there exist integers $N_1, N_2 \geq 1$ such that

$$
|a_n - l| < \frac{\varepsilon}{2(1 + M + |l|)} \quad \forall \text{ integer } n \geq N_1
$$

and

$$
|b_n - m| < \frac{\varepsilon}{2(1 + M + |l|)} \quad \forall \text{ integer } n \geq N_2 .
$$

Thus, for each integer $n \geq \max\{N_1, N_2\}$, $(**)$ implies

$$
\begin{aligned}
|a_n b_n - lm| &< M \frac{\varepsilon}{2(1 + M + |l|)} + |l| \frac{\varepsilon}{2(1 + M + |l|)} \\
&< \frac{\varepsilon}{2} + \frac{\varepsilon}{2} = \varepsilon ,
\end{aligned}
$$

and the proof is complete because we have shown that the definition (ix) is satisfied for the sequence $a_n b_n$ with lm in place of l.

Remark: Notice that if we take $\{a_n\}$ to be the constant sequence $-1, -1, \ldots$ (so that trivially $\lim a_n = -1$) in part (ii) above, we get

$$
\lim(-b_n) = -\lim b_n .
$$

Hence, using part (i) with $\{-b_n\}$ in place of b_n we conclude

$$
\lim(a_n - b_n) = \lim a_n - \lim b_n .
$$

By similar argument (i), (ii), imply that

$$
\lim(\alpha a_n + \beta b_n) = \alpha \lim a_n + \beta \lim b_n .
$$

For any $\alpha, \beta \in \mathbb{R}$ (provided $\lim a_n$, $\lim b_n$ both exist).

The following definition of *subsequence* is important.

Definition: Given a sequence $a_1, a_2, \ldots$, we say $a_{n_1}, a_{n_2}, \ldots$ is a *subsequence* of $\{a_n\}$ if $n_1, n_2, \ldots$ are integers with $1 \leq n_1 < n_2 < n_3 < \ldots$. (Note the *strict* inequalities.)

Thus, $\{b_n\}_{n=1,2,\ldots}$ is a subsequence of $\{a_n\}_{n=1,2,\ldots}$ if and only if *for each* $j \geq 1$ *both* the following hold:

(i) b_j is one of the terms of the sequence $a_1, a_2, \ldots$, and

(ii)   the real number b_{j+1} appears later than the real number b_j in the sequence $a_1, a_2, \ldots$

Thus, $\{a_{2n}\}_{n=1,2,\ldots} = a_2, a_4, a_6 \ldots$ is a subsequence of $\{a_n\}_{n=1,2,\ldots}$ and $\frac{1}{2}, \frac{1}{3}, \frac{1}{4}, \frac{1}{5} \ldots$ is a subsequence of $\{\frac{1}{n}\}_{n=1,2,\ldots}$ but $\frac{1}{3}, \frac{1}{2}, \frac{1}{4}, \frac{1}{5}, \ldots$ is *not* a subsequence of $\{\frac{1}{n}\}$.

Theorem 2.4. (Bolzano-Weierstrass Theorem.) *Any bounded sequence $\{a_n\}$ has a convergent subsequence.*

Proof: ("Method of bisection") Since $\{a_n\}$ is bounded we can find upper and lower bounds, respectively. Thus, $\exists\, c < d$ such that

(∗) $c \leq a_k \leq d \ \forall$ integer $k \geq 1$.

Now subdivide the interval $[c, d]$ into the two half intervals $[c, \frac{c+d}{2}]$, $[\frac{c+d}{2}, d]$. By (∗), at least one of these (call it I_1 say) has the property that there are *infinitely many* integers k with $a_k \in I_1$. Similarly, we can divide I_1 into two equal subintervals (each of length $\frac{d-c}{4}$), at least one of which (call it I_2) has the property that $a_k \in I_2$ for infinitely many integers k. Proceeding in this way we get a sequence of intervals $\{I_n\}_{n=1,2\ldots}$ with $I_n = [c_n, d_n]$ and with the properties that, for each integer $n \geq 1$,

(1)   length $I_n \equiv d_n - c_n = \dfrac{d-c}{2^n}$

(2) $[c_{n+1}, d_{n+1}] \subset [c_n, d_n] \subset [c, d]$

(3) $a_k \in [c_n, d_n]$ for infinitely many integers $k \geq 1$.

(Notice that (3) says $c_n \leq a_k \leq d_n$ for infinitely many integers $k \geq 1$.)

Using properties (1), (2), (3), we proceed to prove the existence of a convergent subsequence as follows.

Select any integer $k_1 \geq 1$ such that $a_{k_1} \in [c_1, d_1]$ (which we can do by (3)). Then select any integer $k_2 > k_1$ with $a_{k_2} \in [c_2, d_2]$. Such k_2 of course exists by (3). Continuing in this way we get integers $1 \leq k_1 < k_2 \ldots$ such that $a_{k_n} \in [c_n, d_n]$ for each integer $n \geq 1$. That is,

(4) $c_n \leq a_{k_n} \leq d_n \ \forall$ integer $n \geq 1$.

On the other hand, by (1), (2) we have

(5) $c \leq c_n \leq c_{n+1} < d_{n+1} \leq d_n \leq d \ \forall$ integer $n \geq 1$.

Notice that (5), in particular, guarantees that $\{c_n\}$, $\{d_n\}$ are *bounded* sequences which are, respectively, *increasing* and *decreasing*, hence by Thm. 2.1 are *convergent*. On the other hand, (1) says $d_n - c_n = \frac{d-c}{2^n}$ ($\to 0$ as $n \to \infty$), hence $\lim c_n = \lim d_n (= \ell$ say). Then by (4) and the Sandwich Theorem (see Exercise 2.5 below) we see that $\{a_{k_n}\}_{n=1,2,\ldots}$ also has limit ℓ.

LECTURE 2 PROBLEMS

2.1   Use the Archimedean property of the reals (Lem. 1.1 of Lecture 1) to prove rigorously that $\lim \frac{1}{n} = 0$.

2.2   Prove part (i) of Thm. 2.1.

Hint: Let $\alpha = \sup S$, and show first that for each $\varepsilon > 0$, $\exists$ an integer $N \geq 1$ such that $a_N > \alpha - \varepsilon$.

2.3   Using the definition (ix) on p. 92 prove that a sequence $\{a_n\}$ cannot have more than one limit.

2.4   If $\{a_n\}$, $\{b_n\}$ are given convergent sequences and $a_n \leq b_n$ $\forall n \geq 1$, prove $\lim a_n \leq \lim b_n$.

Hint: $\lim(a_n - b_n) = \lim a_n - \lim b_n$, so it suffices to prove that $\lim c_n \leq 0$ whenever $\{c_n\}$ is convergent with $c_n \leq 0$ $\forall n$.

2.5   ("Sandwich Theorem.") If $\{a_n\}$, $\{b_n\}$ are given convergent sequences with $\lim a_n = \lim b_n$, and if $\{c_n\}$ is any sequence such that $a_n \leq c_n \leq b_n$ $\forall n \geq 1$, prove that $\{c_n\}$ is convergent and $\lim c_n = \lim a_n (= \lim b_n)$ (Hint: Let $l = \lim a_n = \lim b_n$ and use the definition (ix) on p. 92.)

2.6   If k is a fixed positive integer and if $\{a_n\}$ is any sequence such that $\frac{1}{n^k} \leq a_n \leq n^k$ $\forall n \geq 1$, prove that $\lim a_n^{1/n} = 1$. (Hint: use 2.5 and the standard limit result that $\lim n^{\frac{1}{n}} = 1$.)

LECTURE 3: CONTINUOUS FUNCTIONS

Here we shall mainly be interested in real valued functions for some closed interval $[a, b]$; thus $f : [a, b] \to \mathbb{R}$. (This is reasonable notation, since for each $x \in [a, b]$, f assigns a value $f(x) \in \mathbb{R}$.) First we recall the definition of *continuity* of such a function.

Definition 1. $f : [a, b] \to \mathbb{R}$ is said to be *continuous at the point* $c \in [a, b]$, if for each $\varepsilon > 0$ there is a $\delta > 0$ such that

$$x \in [a, b] \text{ with } |x - c| < \delta \Rightarrow |f(x) - f(c)| < \varepsilon .$$

Definition 2. We say $f : [a, b] \to \mathbb{R}$ is continuous if f is continuous at *each* point $c \in [a, b]$.

We want to prove the important theorem that such a continuous function attains both its maximum and minimum values on $[a, b]$. We first make the terminology precise.

Terminology: If $f : [a, b] \to \mathbb{R}$, then:

(1) f is said to *attain its maximum* at a point $c \in [a, b]$ if $f(x) \le f(c) \, \forall x \in [a, b]$;

(2) f is said to *attain its minimum* at a point $c \in [a, b]$ if $f(x) \ge f(c) \, \forall x \in [a, b]$.

We shall also need the following lemma, which is of independent importance.

Lemma 3.1. *If* $a_n \in [a, b] \, \forall n \ge 1$ *and if* $\lim a_n = c \in [a, b]$ *and if* $f : [a, b] \to \mathbb{R}$ *is continuous at* c, *then*

$$\lim f(a_n) = f(c) ,$$

i.e., the sequence $\{f(a_n)\}_{n=1,2\ldots}$ *converges to* $f(c)$.

Proof: Let $\varepsilon > 0$. By Def. 1 above, $\exists \, \delta > 0$ such that

$(*)$ $\qquad\qquad |f(x) - f(c)| < \varepsilon$ whenever $x \in [a, b]$ with $|x - c| < \delta$.

On the other hand, by the definition of $\lim a_n = c$, with δ in place of ε (i.e., we use the definition (ix) on p. 92 of Lecture 2 with δ in place of ε) we can find an integer $N \ge 1$ such that

$$|a_n - c| < \delta \text{ whenever } n \ge N .$$

Then, (since $a_n \in [a, b] \, \forall n$) $(*)$ tells us that

$$|f(a_n) - f(c)| < \varepsilon \, \forall n \ge N .$$

Theorem 3.2. *If* $f : [a, b] \to \mathbb{R}$ *is continuous, then* f *is bounded and* $\exists$ *points* $c_1, c_2, \in [a, b]$ *such that* f *attains its maximum at the point* c_1 *and its minimum at the point* c_2; *that is,* $f(c) \le f(x) \le f(c_2) \, \forall x \in [a, b]$.

Proof: It is enough to prove that f bounded above and that there is a point $c_1 \in [a, b]$ such that f attains its maximum at c_1, because we can get the rest of the theorem by applying this results to $-f$.

To prove f is bounded above we argue by contradiction. If f is not bounded above, then for each integer $n \geq 1$ we can find a point $x_n \in [a, b]$ such that $f(x) \geq n$. Since x_n is a bounded sequence, by the Bolzano-Weierstrass Theorem (Thm. 2.4 of Lecture 2) we can find a convergent subsequence $x_{n_1}, x_{n_2}, \ldots$ Let $c = \lim x_{n_j}$

Of course, since $a \leq x_{n_j} \leq b \,\forall j$, we must have $c \in [a, b]$. Also, since $1 \leq n_1 < n_2 < \ldots$ (and since $n_1, n_2 \ldots$ are integers), we must have $n_{j+1} \geq n_j + 1$, hence by induction on j

(1) $n_j \geq j \quad \forall \text{ integer } j \geq 1$.

Now since $x_{n_j} \in [a, b]$ and $\lim x_{n_j} = c \in [a, b]$ we have by Lem. 3.1 that $\lim f(x_{n_j}) = f(c)$. Thus, $f(x_{n_j})_{j=1,2,\ldots}$ is *convergent*, hence *bounded* by Thm. 2.2. But by construction $f(x_{n_j}) \geq n_j (\geq j$ by (1)), hence $f(x_{n_j})_{j=1,2\ldots}$ is not bounded, a contradiction. This completes the proof that f is bounded above.

We now want to prove f attains its maximum value at some point $c_1 \in [a, b]$. Let $S = \{f(x) : x \in [a, b]\}$. We just proved above that S is bounded above, hence (since it is non empty by definition) S has a *least upper bound* which we denote by M. We claim that for each integer $n \geq 1$ there is a point $x_n \in [a, b]$ such that $f(x_n) > M - \frac{1}{n}$. Indeed, *otherwise $M - \frac{1}{n}$ would be an upper bound for S*, contradicting the fact that M was chosen to be the *least* upper bound. Again we can use the Bolzano-Weierstrass Theorem to find a convergent subsequence $x_{n_1}, x_{n_2} \ldots$ and again (1) holds. Let $c_1 = \lim x_{n_j}$. By Lem. 3.1 again we have

(2) $f(c_1) = \lim f(x_{n_j})$.

However, by construction we have

$$M \geq f(x_{n_j}) > M - \frac{1}{n_j} \geq M - \frac{1}{j} \quad \text{(by (1))}.$$

And hence by the Sandwich Theorem (Exercise 2.4 of Lecture 2) we have $\lim f(x_{n_j}) = M$. By (2) this gives $f(c_1) = M$. But M is an upper bound for $S = \{f(x) : x \in [a, b]\}$, hence we have $f(x) \leq f(c_1) \,\forall x \in [a, b]$, as required.

An important consequence of the above theorem is the following.

Lemma (Rolle's Theorem): *If $f : [a, b] \to \mathbb{R}$ is continuous, if $f(a) = f(b) = 0$ and if f is differentiable at each point of (a, b), then there is a point $c \in (a, b)$ with $f'(c) = 0$.*

Proof: If f is identically zero then $f'(c) = 0$ for every point $c \in (a, b)$, so assume f is not identically zero. Without loss of generality we may assume $f(x) > 0$ for some $x \in (a, b)$ (otherwise this property holds with $-f$ in place of f). Then max f (which exists by Thm. 3.2) is positive and is attained at some point $c \in (a, b)$. We claim that $f'(c) = 0$. Indeed, $f'(c) = \lim_{x \to c} \frac{f(x) - f(c)}{x - c} = \lim_{x \to c_+} \frac{f(x) - f(c)}{x - c} = \lim_{x \to c_-} \frac{f(x) - f(c)}{x - c}$ and the latter 2 limits are, respectively, ≤ 0 and ≥ 0. But they are equal, hence they must both be zero.

Corollary (Mean Value Theorem): *If $f : [a, b] \to \mathbb{R}$ is continuous and f is differentiable at each point of (a, b), then there is some point $c \in (a, b)$ with $f'(c) = \frac{f(b) - f(a)}{b - a}$.*

Proof: Apply Rolle's Theorem to the function $\widetilde{f}(x) = f(x) - f(a) - \frac{f(b) - f(a)}{b - a}(x - a)$.

LECTURE 3 PROBLEMS

3.1   Give an example of a bounded function $f : [0, 1] \to \mathbb{R}$ such that f is continuous at each point $c \in [0, 1]$ except at $c = 0$, and such that f attains neither a maximum nor a minimum value.

3.2   Prove carefully (using the definition of continuity on p. 96) that the function $f : [-1, 1] \to \mathbb{R}$ defined by

$$f(x) = \begin{cases} +1 & \text{if} \quad 0 < x \leq 1 \\ 0 & \text{if} -1 \leq x \leq 0 \end{cases}$$

is not continuous at $x = 0$. (Hint: Show the definition fails with, e.g., $\varepsilon = \frac{1}{2}$.)

3.3   Let $f : [a, b] \to \mathbb{R}$ be continuous, and let $|f| : [a, b] \to \mathbb{R}$ be defined by $|f|(x) = |f(x)|$. Prove that $|f|$ is continuous.

3.4   Suppose $f : [a, b] \to \mathbb{R}$ and $c \in [a, b]$ are given, and suppose that $\lim f(x_n) = f(c)$ for all sequences $\{x_n\}_{n=1, 2, \ldots} \subset [a, b]$ with $\lim x_n = c$. Prove that f is continuous at c. (Hint: If not, $\exists \varepsilon > 0$ such that $(*)$ on p. 96 fails for each $\delta > 0$; in particular, $\exists \varepsilon > 0$ such that $(*)$ fails for $\delta = \frac{1}{n}$ $\forall$ integer $n \geq 1$.)

3.5   If $f : [0, 1] \to \mathbb{R}$ is defined by

$$f(x) = \begin{cases} 1 \text{ if } x \in [0, 1] \text{ is a rational number} \\ 0 \text{ if } x \in [0, 1] \text{ is not rational .} \end{cases}$$

Prove that f is continuous at no point of $[0, 1]$.

Hint: Recall that any interval $(c, d) \in \mathbb{R}$ (with $c < d$) contains both rational and irrational numbers.

3.6   Suppose $f : (0, 1) \to \mathbb{R}$ is defined by $f(x) = 0$ if $x \in (0, 1)$ is not a rational number, and $f(x) = 1/q$ if $x \in (0, 1)$ can be written in the form $x = \frac{p}{q}$ with p, q positive integers without common factors. Prove that f is continuous at each irrational value $x \in (0, 1)$.

Hint: First note that for a given $\varepsilon > 0$ there are at most finitely many positive integers q with $\frac{1}{q} \geq \varepsilon$.

3.7   Suppose $f : [0, 1] \to \mathbb{R}$ is continuous, and $f(x) = 0$ for each rational point $x \in [0, 1]$. Prove $f(x) = 0$ for all $x \in [0, 1]$.

3.8   If $f : \mathbb{R} \to \mathbb{R}$ is continuous at each point of $\mathbb{R}$, and if $f(x + y) = f(x) + f(y)\, \forall x, y \in \mathbb{R}$, prove $\exists$ a constant a such that $f(x) = ax\, \forall x \in \mathbb{R}$. Show by example that the result is false if we drop the requirement that f be continuous.

LECTURE 4: SERIES OF REAL NUMBERS

Consider the series

$$a_1 + a_2 + \cdots + a_n + \ldots$$

(usually written with summation notation as $\sum_{n=1}^{\infty} a_n$), where $a_1, a_2, \ldots$ is a given sequence of real numbers. a_n is called the n-th term of the series. The sum of the first n terms is

$$s_n = \sum_{k=1}^{n} a_k \; ;$$

s_n is called the n-th partial sum of the series. If

$$s_n \to s$$

(i.e., if $\lim s_n = s$) for some $s \in \mathbb{R}$, then we say the series *converges*, and *has sum s*. Also, in this case we write

$$s = \sum_{n=1}^{\infty} a_n \; .$$

If s_n does not converge, then we say the series *diverges*.

Example: If $a \in \mathbb{R}$ is given, then the series $1 + a + a^2 + \ldots$ (i.e., the geometric series) has nth partial sum

$$s_n = 1 + a + \cdots + a^{n-1} = \begin{cases} n & \text{if } a = 1 \\ \frac{1-a^n}{1-a} & \text{if } a \neq 1 \; . \end{cases}$$

Using the fact that $a^n \to 0$ if $|a| < 1$, we thus see that the series converges and has sum $\frac{1}{1-a}$ if $|a| < 1$, whereas the series diverges for $|a| \geq 1$. (Indeed $\{s_n\}$ is *unbounded* if $|a| > 1$ or $a = 1$, and if $a = -1$, $\{s_n\}_{n=1,2,\ldots} = 1, 0, 1, 0, \ldots$.)

The following simple lemma is of key importance.

Lemma 4.1. If $\sum_{n=1}^{\infty} a_n$ converges, then $\lim a_n = 0$.

Note: The *converse* is not true. For example, we check below that $\sum_{n=1}^{\infty} \frac{1}{n}$ does *not* converge, but its nth term is $\frac{1}{n}$, which *does* converge to zero.

Proof of Lemma 4.1: Let $s = \lim s_n$. Then of course we also have $s = \lim s_{n+1}$. But, $s_{n+1} - s_n = (a_1 + a_2 + \cdots + a_{n+1}) - (a_1 + a_2 + \cdots a_n) = a_{n+1}$, hence (see the remark following Thm. 2.3 of Lecture 2) we have $\lim a_{n+1} = \lim s_{n+1} - \lim s_n = s - s = 0$. i.e., $\lim a_n = 0$.

The following lemma is of theoretical and practical importance.

Lemma 4.2. *If $\sum_{n=1}^{\infty} a_n$, $\sum_{n=1}^{\infty} b_n$ both converge, and have sum s, t, respectively, and if α, β are real numbers, then $\sum_{n=1}^{\infty} (\alpha a_n + \beta b_n)$ also converges and has sum $\alpha s + \beta t$, i.e., $\sum_{n=1}^{\infty} (\alpha a_n + \beta b_n) = \alpha \sum_{n=1}^{\infty} a_n + \beta \sum_{n=1}^{\infty} b_n$ if both $\sum_{n=1}^{\infty} a_n$, $\sum_{n=1}^{\infty} b_n$ converge.*

Proof: Let $s_n = \sum_{k=1}^{n} a_n$, $t_n = \sum_{k=1}^{n} b_k$. We are given $s_n \to s$ and $t_n \to t$, then $\alpha s_n + \beta t_n \to \alpha s + \beta t$ (see the remarks following Thm. 2.3 of Lecture 2). But $\alpha s_n + \beta t_n = \alpha \sum_{k=1}^{n} a_n + \beta \sum_{k=1}^{n} b_k = \sum_{k=1}^{n} (\alpha a_k + \beta b_k)$, which is the nth partial sum of $\sum_{n=1}^{\infty} (\alpha a_n + \beta b_n)$.

There is a very convenient criteria for checking convergence in case all the terms are *nonnegative*. Indeed, in this case

$$s_{n+1} - s_n = a_{n+1} \geq 0 \, \forall n \geq 1 \, ,$$

hence the sequence $\{s_n\}$ is *increasing* if $a_n \geq 0$. Thus, by Thm. 2.1(i) of Lecture 2 we see that $\{s_n\}$ converges if and only if it is *bounded*. That is, we have proved:

Lemma 4.3. *If each term of $\sum_{n=1}^{\infty} a_n$ is nonnegative (i.e., if $a_n \geq 0 \, \forall n$) then the series converges if and only if the sequence of partial sums (i.e., $\{s_n\}_{n=1,2,...}$) is bounded.*

Example. Using the above criteria, we can discuss convergence of $\sum_{n=1}^{\infty} \frac{1}{n^p}$ where $p > 0$ is given. The nth partial sum in this case is

$$s_n = \sum_{k=1}^{n} \frac{1}{k^p} \, .$$

Since $\frac{1}{x^p}$ is a decreasing function of x for $x > 0$, we have, for each integer $k \geq 1$,

$$\frac{1}{(k+1)^p} \leq \frac{1}{x^p} \leq \frac{1}{k^p} \quad \forall x \in [k, k+1] \, .$$

Integrating, this gives

$$\frac{1}{(k+1)^p} \equiv \int_{k}^{k+1} \frac{1}{(k+1)^p} \leq \int_{k}^{k+1} \frac{1}{x^p} \, dx \leq \int_{k}^{k+1} \frac{1}{k^p} \, dx \equiv \frac{1}{k^p} \, ,$$

so if we sum from $k = 1$ to n, we get

$$\sum_{k=1}^{n} \frac{1}{(k+1)^p} \leq \int_{1}^{n+1} \frac{1}{x^p} \, dx \leq \sum_{k=1}^{n} \frac{1}{k^p} \, .$$

That is,

$$s_{n+1} - 1 \leq \int_{1}^{n+1} \frac{1}{x^p} \, dx \leq s_n \quad \forall n \geq 1 \, .$$

But

$$\int_{1}^{n+1} \frac{1}{x^p} \, dx = \begin{cases} \log(n+1) & \text{if } p = 1 \\ \frac{(n+1)^{1-p} - 1}{1-p} & \text{if } p \neq 1 \, . \end{cases}$$

Thus, we see that $\{s_n\}$ is *unbounded* if $p \leq 1$ and bounded for $p > 1$, hence from Lem. 4.2 we conclude

$$\sum_{n=1}^{\infty} \frac{1}{n^p} \begin{cases} \text{converges} & p > 1 \\ \text{diverges} & p \leq 1 \, . \end{cases}$$

Remark. The above method can be modified to discuss convergence of other series. See Exercise 4.6 below.

Theorem 4.4. If $\sum_{n=1}^{\infty} |a_n|$ converges, then $\sum_{n=1}^{\infty} a_n$ converges.

Terminology: If $\sum_{n=1}^{\infty} |a_n|$ converges, then we say that $\sum_{n=1}^{\infty} a_n$ is *absolutely convergent*. Thus, with this terminology, the above theorem just says "absolute convergence $\Rightarrow$ convergence."

Proof of Theorem 4.4: Let $s_n = \sum_{k=1}^{n} a_k$, $t_n = \sum_{k=1}^{n} |a_k|$. Then we are given $t_n \to t$ for some $t \in \mathbb{R}$.

For each integer $n \geq 1$, let

$$
p_n = \begin{cases} a_n & \text{if } a_n \geq 0 \\ 0 & \text{if } a_n < 0 \end{cases}
$$

$$
q_n = \begin{cases} -a_n & \text{if } a_n \leq 0 \\ 0 & \text{if } a_n > 0 \,, \end{cases}
$$

and let $s_n^+ = \sum_{k=1}^{n} p_n$, $s_n^- = \sum_{k=1}^{n} q_n$. Notice that for each $n \geq 1$ we then have

$$
\begin{aligned}
a_n &= p_n - q_n, & s_n &= s_n^+ - s_n^- \\
|a_n| &= p_n + q_n, & t_n &= s_n^+ + s_n^- \,,
\end{aligned}
$$

and $p_n, q_n \geq 0$. Also,

$$
0 \leq s_n^+ \leq t_n \leq t \text{ and } 0 \leq s_n^- \leq t_n \leq t \,.
$$

Hence, we have shown that $\sum_{n=1}^{\infty} p_n$, $\sum_{n=1}^{\infty} q_n$ have bounded partial sums. Hence, by Lem. 4.3, both $\sum_{n=1}^{\infty} p_n$, $\sum_{n=1}^{\infty} q_n$ converge. But then (by Lem. 4.2) $\sum_{n=1}^{\infty} (p_n - q_n)$ converges, i.e., $\sum_{n=1}^{\infty} a_n$ converges.

Rearrangement of series: We want to show that the terms of an absolutely convergent series can be rearranged in an arbitrary way without changing the sum. First we make the definition clear.

Definition: Let $j_1, j_2, \ldots$ be any sequence of positive integers in which every positive integer appears once and only once (i.e., the mapping $n \to j_n$ is a $1 : 1$ mapping of the positive integers onto the positive integers). Then the series $\sum_{n=1}^{\infty} a_{j_n}$ is said to be a *rearrangement* of the series $\sum_{n=1}^{\infty} a_n$.

Theorem 4.5. *If* $\sum_{n=1}^{\infty} a_n$ *is absolutely convergent, then any rearrangement* $\sum_{n=1}^{\infty} a_{j_n}$ *converges, and has the same sum as* $\sum_{n=1}^{\infty} a_n$.

Proof: We give the proof when $a_n \geq 0 \; \forall n$ (in which case "absolute convergence" just means "convergence"). This extension to the general case is left as a exercise. (See Exercise 4.8 below.)

Hence, assume $\sum_{n=1}^{\infty} a_n$ converges, and $a_n \geq 0 \; \forall$ integer $n \geq 1$, and let $\sum_{n=1}^{\infty} a_{j_n}$ be any rearrangement. For each $n \geq 1$, let

$$
P(n) = \max\{j_1, \ldots, j_n\} \,.
$$

So that

$$\{j_1, \ldots, j_n\} \subset \{1, \ldots P(n)\} \text{ and hence (since } a_n \geq 0 \ \forall k)$$

$$a_{j_1} + a_{j_2} + \cdots + a_{j_n} \leq a_1 + \cdots + a_{P(n)} \leq s,$$

where $s = \sum_{n=1}^{\infty} a_n$. Thus, we have shown that the partial sums of $\sum_{n=1}^{\infty} a_{j_n}$ are bounded above by s, hence by Lem. 4.3, $\sum_{n=1}^{\infty} a_{j_n}$ converges, and has sum t satisfying $t \leq s$. But $\sum_{n=1}^{\infty} a_n$ is a rearrangement of $\sum_{n=1}^{\infty} a_{j_n}$ (using the rearrangement given by the inverse mapping $j_n \to n$), and hence by the same argument we also have $s \leq t$. Hence, $s = t$ as required.

LECTURE 4 PROBLEMS

The first few problems give various criteria to test for *absolute convergence* (and also, in some cases, for testing for *divergence*).

4.1 (i) (Comparison test.) If $\sum_{n=1}^{\infty} a_n$, $\sum_{n=1}^{\infty} b_n$ are given series and if $|a_n| \leq |b_n| \ \forall n \geq 1$, prove $\sum_{n=1}^{\infty} b_n$ absolutely convergent $\Rightarrow \sum_{n=1}^{\infty} a_n$ absolutely convergent.

(ii) Use this to discuss convergence of:

  (a) $\sum_{n=1}^{\infty} \frac{\sin n}{n^2}$

  (b) $\sum_{n=1}^{\infty} \frac{\sin(\frac{1}{n})}{n}$.

4.2 (Comparison test for *divergence*.) If $\sum_{n=1}^{\infty} a_n$, $\sum_{n=1}^{\infty} b_n$ are given series with nonnegative terms, and if $a_n \geq b_n \ \forall n \geq 1$, prove $\sum_{n=1}^{\infty} b_n$ diverges $\Rightarrow \sum_{n=1}^{\infty} a_n$ diverges.

4.3 (i) (Ratio Test.) If $a_n \neq 0 \ \forall n$, if $\lambda \in (0, 1)$ and if there is an integer $N \geq 1$ such that $\frac{|a_{n+1}|}{|a_n|} \leq \lambda \ \forall n \geq N$, prove that $\sum_{n=1}^{\infty} a_n$ is absolutely convergent. (Hint: First use induction to show that $|a_n| \leq \lambda^{n-N} \ \forall n \geq N$.)

(ii) (Ratio test for divergence.) If $a_n \neq 0 \ \forall n$ and if $\exists$ an integer $N \geq 1$ such that $\frac{|a_{n+1}|}{|a_n|} \geq 1 \ \forall n \geq N$ then prove $\sum_{n=1}^{\infty} a_n$ diverges.

4.4 (Cauchy root test.) (i) Suppose $\exists \lambda \in (0, 1)$ and an integer $N \geq 1$ such that $|a_n|^{\frac{1}{n}} \leq \lambda \ \forall n \geq N$. Prove that $\sum_{n=1}^{\infty} a_n$ converges.

(ii) Use the Cauchy root test to discuss convergence of $\sum_{n=1}^{\infty} n^2 x^n$. (Here $x \in \mathbb{R}$ is given—consider the possibilities $|x| < 1$, $|x| > 1$, $|x| = 1$.)

4.5 Suppose $a_n \geq 0 \ \forall n \geq 1$ and $\sum_{n=1}^{\infty} a_n$ diverges. Prove

(i) $\sum_{n=1}^{\infty} \frac{a_n}{1+a_n}$ diverges

(ii) $\sum_{n=1}^{\infty} \frac{a_n}{1+n^2 a_n}$ converges.

4.6 (Integral Test.) If $f : [1, \infty) \to \mathbb{R}$ is positive and continuous at each point of $[1, \infty)$, and if f is *decreasing*, i.e., $x < y \Rightarrow f(y) \leq f(x)$, prove using a modification of the argument on pp. 100–102 that $\sum_{n=1}^{\infty} f(n)$ converges if and only if $\{\int_1^n f(x)\,dx\}_{n=1,2,\ldots}$ is *bounded*.

4.7   Using the integral test (in Exercise 4.6 above) to discuss convergence of

(i) $\sum_{n=2}^{\infty} \frac{1}{n \log n}$

(ii) $\sum_{n=2}^{\infty} \frac{1}{n(\log n)^{1+\varepsilon}}$, where $\varepsilon > 0$ is a given constant.

4.8   Complete the proof of Thm. 4.5 (i.e., discuss the general case when $\sum |a_n|$ converges). (Hint: The theorem has already been established for series of *nonnegative* terms; use p_n, q_n as in Thm. 4.4.)

LECTURE 5: POWER SERIES

A power series is a series of the form $\sum_{n=0}^{\infty} a_n x^n$, where $a_0, a_1, \ldots$ are given real numbers and x is a real variable. Here we use the standard convention that $x^0 = 1$, so the first term $a_0 x^0$ just means a_0.

Notice that for $x = 0$ the series trivially converges and its sum is a_0.

The following lemma describes the key convergence property of such series.

Lemma 1. *If the series $\sum_{n=0}^{\infty} a_n x^n$ converges for some value $x = c$, then the series converges absolutely for every x with $|x| < |c|$.*

Proof: $\sum_{n=0}^{\infty} a_n c^n$ converges $\Rightarrow \lim a_n c^n = 0 \Rightarrow \{a_n c^n\}_{n=1,2,\ldots}$ is a bounded sequence. That is, there is a fixed constant $M > 0$ such that $|a_n c^n| \le M \ \forall n = 0, 1, \ldots$, and so $|x| < |c| \Rightarrow$, for any $j = 0, 1, \ldots$,

$$|a_j x^j| = |a_j c^j| \left| \frac{x^j}{c^j} \right| \le M \left| \frac{x^j}{c^j} \right| = M \left(\frac{|x|}{|c|} \right)^j = M r^j, \quad r = \frac{|x|}{|c|} < 1,$$

and hence

$$\sum_{j=0}^{n-1} |a_j x^j| \le M \sum_{j=0}^{n-1} r^j = M \frac{1 - r^n}{1 - r} \le \frac{M}{1 - r}, \quad n = 1, 2, \ldots .$$

Thus, the series $\sum_{n=0}^{\infty} |a_n x^n|$ is convergent, because it has nonnegative terms and we've shown its partial sums are bounded. This completes the proof.

We can now directly apply the above lemma to establish the following basic property of power series.

Theorem 5.1. *For any given power series $\sum_{n=0}^{\infty} a_n x^n$, exactly one of the following 3 possibilities holds:*

(i) *the series diverges $\forall x \ne 0$, or*

(ii) *the series converges absolutely $\forall x \in \mathbb{R}$, or*

(iii) *$\exists \rho > 0$ such that the series converges absolutely $\forall x$ with $|x| < \rho$, and diverges $\forall x$ with $|x| > \rho$.*

Terminology: If (iii) holds, the number ρ is called the *radius of convergence* and the interval $(-\rho, \rho)$ is called the *interval of convergence*. If (i) holds we say the radius of convergence is zero, and if (ii) holds we say the radius of convergence $= \infty$.

Note: The theorem says nothing about what happens at $x = \pm \rho$ in case (iii).

Proof of Theorem 5.1: Consider the set $S \subset \mathbb{R}$ defined by

$$S = \{ |x| : \sum_{n=0}^{\infty} a_n x^n \text{ converges} \} .$$

Notice that we always have $0 \in S$, so S is nonempty. If $S = \{0\}$ then case (i) holds, so we can assume S contains at least one c with $c \ne 0$. If S is not bounded then by Lem. 1 we have that $\sum_{n=0}^{\infty} a_n x^n$ is absolutely convergent (A.C.) on $(-R, R)$ for each $R > 0$, and hence (ii) holds. If

$S \neq \{0\}$ is bounded then $R = \sup S$ exists and is positive. Now for any $x \in (-R, R)$ we have $c \in S$ with $|c| > |x|$ (otherwise $|c| \leq |x|$ for each $c \in S$ meaning that $|x|$ would be an upper bound for S smaller than R, contradicting $R = \sup S$), and hence by Lem. 1 $\sum a_n x^n$ is A.C. So, in fact, $\sum_{n=0}^{\infty} a_n x^n$ is A.C. for each $x \in (-R, R)$. We must of course also have $\sum_{n=0}^{\infty} a_n x^n$ diverges for each x with $|x| > R$ because otherwise we would have x_0 with $|x_0| > R$ and $\sum_{n=0}^{\infty} a_n x_0^n$ convergent, hence $|x_0| \in S$, which contradicts the fact that R is an upper bound for S.

Suppose now that a given power series $\sum_{n=0}^{\infty} a_n x^n$ has radius of convergence $\rho > 0$ (we include here the case $\rho = \infty$, which is to be interpreted to mean that the series converges absolutely for all $x \in \mathbb{R}$).

A reasonable and natural question is whether or not we can also *expand $f(x)$ in terms of powers of $x - \alpha$* for given $\alpha \in (-\rho, \rho)$. The following theorem shows that we can do this for $|x - \alpha| < \rho - |\alpha|$ (and for all x in case $\rho = \infty$).

Theorem 5.2 (Change of Base-Point.) *If $f(x) = \sum_{n=0}^{\infty} a_n x^n$ has radius of convergence $\rho > 0$ or $\rho = \infty$, and if $|\alpha| < \rho$ (and α arbitrary in case $\rho = \infty$), then we can also write*

(*) $f(x) = \sum_{m=0}^{\infty} b_m (x - \alpha)^m$ $\forall x$ *with $|x - \alpha| < \rho - |\alpha|$ (x, α arbitrary if $\rho = \infty$)*,

where $b_m = \sum_{n=m}^{\infty} \binom{n}{m} a_n \alpha^{n-m}$ (so, in particular, $b_0 = f(\alpha)$); part of the conclusion here is that the series for b_m converges, and the series $\sum_{m=0}^{\infty} b_m (x - \alpha)^m$ converges for the stated values of x, α.

Note: The series on the right in (*) is a power series in powers of $x - \alpha$, hence the fact that it converges for $|x - \alpha| < \rho - |\alpha|$ means that it has radius of convergence (as a power series in powers of $x - \alpha$) $\geq \rho - |\alpha|$ (and radius of convergence $= \infty$ in case $\rho = \infty$). Thus, in particular, the series on the right of (*) automatically converges *absolutely* for $|x - \alpha| < \rho - |\alpha|$ by Lem. 1.

Proof of Theorem 5.2: We take any α with $|\alpha| < \rho$ and any x with $|x - \alpha| < \rho - |\alpha|$ (α, x are arbitrary if $\rho = \infty$), and we look at the partial sum $S_N = \sum_{n=0}^{N} a_n x^n$. Since the Binomial Theorem tells us that $x^n = (\alpha + (x - \alpha))^n = \sum_{m=0}^{n} \binom{n}{m} \alpha^{n-m} (x - \alpha)^m$, we see that S_N can be written

$$\sum_{n=0}^{N} a_n x^n = \sum_{n=0}^{N} a_n \sum_{m=0}^{n} \binom{n}{m} \alpha^{n-m} (x - \alpha)^m .$$

Using the interchange of sums formula (see Problem 6.3 below)

$$\sum_{n=0}^{N} \sum_{m=0}^{n} c_{nm} = \sum_{m=0}^{N} \sum_{n=m}^{N} c_{nm} ,$$

this then gives

(1) $$\sum_{n=0}^{N} a_n x^n = \sum_{m=0}^{N} \left(\sum_{n=m}^{N} \binom{n}{m} a_n \alpha^{n-m} \right) (x - \alpha)^m .$$

Now since $\binom{n}{m}^{1/n} \to 1$ as $n \to \infty$ for each fixed m, we see that for any $\varepsilon > 0$ we have N such that $\binom{n}{m} \leq (1 + \varepsilon)^n$ for all $n \geq N$, hence $|a_n \binom{n}{m} x^n| \leq |a_n ((1 + \varepsilon) x)^n|$ $\forall n \geq N$ and hence by the comparison test $\sum_{n=0}^{\infty} \binom{n}{m} a_n x^n$ also converges absolutely for all $x \in (-\rho, \rho)$ (because $|x| < \rho \Rightarrow (1 +$

$\varepsilon)|x| < \rho$ for suitable $\varepsilon > 0$). Thus, since $|\alpha| < \rho$ we have, in particular, that $\sum \binom{n}{m} a_n \alpha^n$ is absolutely convergent and we can substitute $\sum_{n=m}^{N} \binom{n}{m} a_n \alpha^{n-m} = \sum_{n=m}^{\infty} \binom{n}{m} a_n \alpha^{n-m} - \sum_{n=N+1}^{\infty} \binom{n}{m} a_n \alpha^{n-m}$ in (1) above, whence (1) gives

$$\sum_{n=0}^{N} a_n x^n = \sum_{m=0}^{N} \left(\sum_{n=m}^{\infty} \binom{n}{m} a_n \alpha^{n-m} \right)(x-\alpha)^m$$

$$- \sum_{m=0}^{N} \left(\sum_{n=N+1}^{\infty} \binom{n}{m} a_n \alpha^{n-m} \right)(x-\alpha)^m$$

and

$$\left| \sum_{m=0}^{N} \left(\sum_{n=N+1}^{\infty} \binom{n}{m} a_n \alpha^{n-m} \right)(x-\alpha)^m \right| \leq \sum_{m=0}^{N} \left(\sum_{n=N+1}^{\infty} \binom{n}{m} |a_n| |\alpha|^{n-m} |x-\alpha|^m \right)$$

$$\equiv \sum_{n=N+1}^{\infty} \left(\sum_{m=0}^{N} \binom{n}{m} |a_n| |\alpha|^{n-m} |x-\alpha|^m \right)$$

$$\leq \sum_{n=N+1}^{\infty} \left(\sum_{m=0}^{n} \binom{n}{m} |a_n| |\alpha|^{n-m} |x-\alpha|^m \right)$$

$$\equiv \sum_{n=N+1}^{\infty} |a_n| (|\alpha| + |x-\alpha|)^n, \quad |\alpha| + |x-\alpha| < \rho ,$$

where we used the Binomial Theorem again in the last line. Now observe that we have

$$\sum_{n=N+1}^{\infty} |a_n| (|\alpha| + |x-\alpha|)^n = \sum_{n=1}^{\infty} |a_n| (|\alpha| + |x-\alpha|)^n$$
$$- \sum_{n=1}^{N} |a_n| (|\alpha| + |x-\alpha|)^n \to 0 \text{ as } N \to \infty ,$$

so the above shows that $\lim_{N\to\infty} \sum_{m=0}^{N} b_m (x-\alpha)^m$ exists (and is real), i.e., the series $\sum_{m=0}^{\infty} b_m (x-\alpha)^m$ converges for $|x-\alpha| < \rho - |\alpha|$, and that the sum of the series agrees with $\sum_{n=0}^{\infty} a_n x^n$, so the proof of Thm. 5.2 is complete.

LECTURE 5 PROBLEMS

5.1 (i) Suppose the radius of convergence of $\sum_{n=0}^{\infty} a_n x^n$ is 1, and the radius of convergence of $\sum_{n=0}^{\infty} b_n x^n$ is 2. Prove that $\sum_{n=0}^{\infty} (a_n + b_n) x^n$ has radius of convergence 1.

(Hint: Lemma 1 guarantees that $\{a_n x^n\}_{n=1,2,\dots}$ is unbounded if $|x|$ is greater than the radius of convergence of $\sum_{n=0}^{\infty} a_n x^n$.)

(ii) If *both* $\sum_{n=0}^{\infty} a_n x^n$, $\sum_{n=0}^{\infty} b_n x^n$ have radius of convergence = 1, show that

   (a) The radius of convergence of $\sum_{n=0}^{\infty} (a_n + b_n) x^n$ is ≥ 1, and

   (b) For any given number $R > 1$ you can construct examples with radius of convergence of $\sum_{n=0}^{\infty} (a_n + b_n) x^n = R$.

5.2 If $\exists$ constants $c, k > 0$ such that $c^{-1} n^{-k} \leq |a_n| \leq c n^k \; \forall n = 1, 2, \dots$, what can you say about the radius of convergence of $\sum_{n=0}^{\infty} a_n x^n$.

LECTURE 6: TAYLOR SERIES REPRESENTATIONS

The change of base point theorem proved in Lecture 5 is actually quite strong; for example, it makes it almost trivial to check that a power series is differentiable arbitrarily many times inside its interval of convergence (i.e., a power series is "C^∞" in its interval of convergence), and furthermore all the derivatives can be correctly calculated simply by differentiating each term (i.e., "termwise" differentiation is a valid method for computing the derivatives of a power series in its interval of convergence). Specifically, we have:

Theorem 6.1. *Suppose the power series $\sum_{n=0}^\infty a_n x^n$ has radius of convergence $\rho > 0$ (or $\rho = \infty$), and let $f(x) = \sum_{n=0}^\infty a_n x^n$ for $|x| < \rho$. Then all derivatives $f^{(m)}(x)$, $m = 1, 2, \ldots$, exist at every point x with $|x| < \rho$, and, in fact,*

$$f^{(m)}(x) = \sum_{n=m}^\infty n(n-1) \cdots (n-m+1) a_n x^{n-m}, \quad x \in (-\rho, \rho),$$

which says precisely that the derivatives of f can be correctly computed simply by differentiating the series $\sum_{n=0}^\infty a_n x^n$ termwise (because $n(n-1) \cdots (n-m+1) a_n x^{n-m}$ is just the mth derivative of $a_n x^n$); that is, $\frac{d^m}{dx^m} \sum_{n=0}^\infty a_n x^n = \sum_{n=0}^\infty \frac{d^m}{dx^m} a_n x^n$ for $|x| < \rho$.

Proof: It is enough to check the stated result $m = 1$, because then the general result follows directly by induction on m.

The proof for $m = 1$ is an easy consequence of Thm. 5.2 (change of base-point theorem), which tells us that we can write

$$(1) \qquad f(x) - f(\alpha) = \sum_{m=1}^\infty b_m (x - \alpha)^m \equiv (x - \alpha) b_1 + (x - \alpha) \sum_{m=2}^\infty b_m (x - \alpha)^{m-1}$$

for $|x - \alpha| < \rho - |\alpha|$, where $b_m = \sum_{n=m}^\infty \binom{n}{m} a_n \alpha^{n-m}$. That is,

$$(2) \qquad \frac{f(x) - f(\alpha) - b_1 (x - \alpha)}{x - \alpha} = \sum_{m=2}^\infty b_m (x - \alpha)^{m-1}, \quad 0 < |x - \alpha| < \rho - |\alpha|$$

and the expression on the right is a convergent power series with radius of convergence at least $r = \rho - |\alpha| > 0$ and hence is A.C. if $x - \alpha$ is in the interval of convergence $(-r, r)$. In particular, it is A.C. when $|x - \alpha| = r/2$ and hence (2) shows that

$$(3) \qquad \left| \frac{f(x) - f(\alpha)}{x - \alpha} - b_1 \right| \le \sum_{m=2}^\infty |b_m| \, |x - \alpha|^{m-1} \le |x - \alpha| \sum_{m=2}^\infty |b_m| (r/2)^{m-2}$$

for $0 < |x - \alpha| < r/2$ (where $r = \rho - |\alpha| > 0$). Since the right side in (3) $\to 0$ as $x \to \alpha$, this shows that $f'(\alpha)$ exists and is equal to $b_1 = \sum_{n=1}^\infty n a_n \alpha^{n-1}$.

We now turn to the important question of which functions f can be expressed as a power series on some interval. Since we have shown power series are differentiable to all orders in their interval of convergence, a *necessary* condition is clearly that f is differentiable to all orders; however, this is *not* sufficient; see Exercise 6.2 below. To get a reasonable sufficient condition, we need the following theorem.

Theorem 6.2 (Taylor's Theorem.) *Suppose $r > 0$, $\alpha \in \mathbb{R}$ and f is differentiable to order $m + 1$ on the interval $|x - \alpha| < r$. Then $\forall x$ with $|x - \alpha| < r$, $\exists c$ between α and x such that*

$$f(x) = \sum_{n=0}^{m} \frac{f^{(n)}(\alpha)}{n!}(x - \alpha)^n + \frac{f^{(m+1)}(c)}{(m + 1)!}(x - \alpha)^{m+1} .$$

Proof: Fix x with $0 < x - \alpha < r$ (a similar argument holds in case $-r < x - \alpha < 0$), and, for $|t - \alpha| < r$, define

(1) $$g(t) = f(t) - \sum_{n=0}^{m} \frac{f^{(n)}(\alpha)}{n!}(t - \alpha)^n - M(t - \alpha)^{m+1} ,$$

where M (constant) is chosen so that $g(x) = 0$, i.e.,

$$M = \frac{(f(x) - \sum_{n=0}^{m} \frac{f^{(n)}(\alpha)}{n!}(x - \alpha)^n)}{(x - \alpha)^{m+1}} .$$

Notice that, by direct computation in (1),

(2) $$\begin{cases} g^{(n)}(\alpha) & = 0 \,\forall n = 0, \ldots, m \\ g^{(m+1)}(t) & = f^{m+1}(t) - M(m + 1)!, \; |t - \alpha| < r. \end{cases}$$

In particular, since $g(\alpha) = g(x) = 0$, the mean value theorem tells us that there is $c_1 \in (\alpha, x)$ such that $g'(c_1) = 0$. But then $g'(\alpha) = g'(c_1) = 0$, and hence again by the mean value theorem there is a constant $c_2 \in (\alpha, c_1)$ such that $g''(c_2) = 0$.

After $(m + 1)$ such steps we deduce that there is a constant $c_{m+1} \in (\alpha, x)$ such that $g^{(m+1)}(c_{m+1}) = 0$. However, by (2), $g^{(m+1)}(t) = f^{(m+1)}(t) - M(m + 1)!$, hence this gives

$$M = \frac{f^{(m+1)}(c_{m+1})}{(m + 1)!} .$$

In view of our definition of M, this proves the theorem with $c = c_{m+1}$.

Theorem 6.2 gives us a satisfactory sufficient condition for writing f in terms of a power series. Specifically we have:

Theorem 6.3. *If $f(x)$ is differentiable to all orders in $|x - \alpha| < r$, and if there is a constant $C > 0$ such that*

(*) $$\left| \frac{f^{(n)}(x)}{n!} \right| r^n \leq C \;\forall n \geq 0, \; and\, \forall x \; with \; |x - \alpha| < r ,$$

then $\sum_{n=0}^{\infty} \frac{f^{(n)}(\alpha)}{n!}(x - \alpha)^n$ converges, and has sum $f(x)$, for every x with $|x - \alpha| < r$.

Note 1: Whether or not (∗) holds, and whether or not $\sum_{n=0}^{\infty} \frac{f^{(n)}(\alpha)}{n!}(x-\alpha)^n$ converges to $f(x)$, we call the series $\sum_{n=0}^{\infty} \frac{f^{(n)}(\alpha)}{n!}(x-\alpha)^n$ the *Taylor series of f* about α.

Note 2: Even if the Taylor series converges in some interval $x - \alpha < r$, it may fail to have sum $f(x)$. (See Exercise 6.2 below). Of course the above theorem tells us that it *will* have sum $f(x)$ in case the additional condition (∗) holds.

Proof of Theorem 6.3: The condition (∗) guarantees that the term $\frac{f^{(m+1)}(c)}{(m+1)!}(x-\alpha)^{m+1}$ on the right in Thm. 6.2 has absolute value $\leq C\left(\frac{|x-\alpha|}{r}\right)^{m+1}$ and hence Thm. 6.2, with $m = N$, ensures

$$\left| f(x) - \sum_{n=0}^{N} \frac{f^{(n)}(\alpha)}{n!}(x-\alpha)^n \right| \leq C\left(\frac{|x-\alpha|}{r}\right)^{N+1} \to 0 \text{ as } N \to \infty$$

if $|x - \alpha| < r$. Hence,

$$\lim_{N \to \infty} \sum_{n=0}^{N} \frac{f^{(n)}(\alpha)}{n!}(x-\alpha)^n = f(x)$$

whenever $|x - \alpha| < r$, i.e.,

$$\sum_{n=0}^{\infty} \frac{f^{(n)}(\alpha)}{n!}(x-\alpha)^n = f(x), \quad |x - \alpha| < r .$$

LECTURE 6 PROBLEMS

6.1 Prove that the function f, defined by

$$f(x) = \begin{cases} e^{-1/x^2} & \text{if } x \neq 0 \\ 0 & \text{if } x = 0 , \end{cases}$$

is C^∞ on $\mathbb{R}$ and satisfies $f^{(m)}(0) = 0 \, \forall \, m \geq 0$.

Note: This means the Taylor series of $f(x)$ about 0 is zero; i.e., it is an example where the Taylor series converges, but the sum is not $f(x)$.

6.2 If f is as in 6.1, prove that there does not exist any interval $(-\varepsilon, \varepsilon)$ on which f is represented by a power series; that is, there cannot be a power series $\sum_{n=0}^{\infty} a_n x^n$ such that $f(x) = \sum_{n=0}^{\infty} a_n x^n$ for all $x \in (-\varepsilon, \varepsilon)$.

6.3 Let b_{nm} be arbitrary real numbers $0 \leq n \leq N, 0 \leq m \leq n$. Prove

$$\sum_{n=0}^{N} \sum_{m=0}^{n} b_{nm} = \sum_{m=0}^{N} \sum_{n=m}^{N} b_{nm} .$$

Hint: Define $\tilde{b}_{nm} = \begin{cases} b_{nm} & \text{if } m \leq n \\ 0 & \text{if } n < m \leq N. \end{cases}$

6.4 Find the Taylor series about $x = 0$ of the following functions; in each case prove that the series converges to the function in the indicated interval.

(i) $\frac{1}{1-x^2}$, $|x| < 1$ (Hint: $\frac{1}{1-y} = 1 + y + y^2 \ldots$, $|y| < 1$).

(ii) $\log(1 + x)$, $|x| < 1$

(iii) e^x, $x \in \mathbb{R}$

(iv) e^{x^2}, $x \in \mathbb{R}$ (Hint: set $y = x^2$).

6.5 (The analytic definition of the functions cos x, sin x, and the number π.)

Let $\sin x$, $\cos x$ be *defined* by $\sin x = \sum_{n=0}^{\infty}(-1)^n \frac{x^{2n+1}}{(2n+1)!}$, $\cos x = \sum_{n=0}^{\infty}(-1)^n \frac{x^{2n}}{(2n)!}$. For convenience of notation, write $C(x) = \cos x$, $S(x) = \sin x$. Prove:

(i) The series defining $S(x)$, $C(x)$ both have radius of convergence ∞, and $S'(x) \equiv C(x)$, $C'(x) \equiv -S(x)$ for all $x \in \mathbb{R}$.

(ii) $S^2(x) + C^2(x) \equiv 1$ for all $x \in \mathbb{R}$. Hint: Differentiate and use (i).

(iii) sin, cos (as defined above) are the unique functions S, C on $\mathbb{R}$ with the properties (a) $S(0) = 0$, $C(0) = 1$ and (b) $S'(x) = C(x)$, $C'(x) = -S(x)$ for all $x \in \mathbb{R}$. Hint: Thus, you have to show that $\widetilde{S} = S$ and $\widetilde{C} = C$ assuming that properties (a),(b) hold with $\widetilde{S}$, $\widetilde{C}$ in place of S, C, respectively; show that Thm. 6.3 is applicable.

(iv) $C(2) < 0$ and hence there is a $p \in (0, 2)$ such that $C(p) = 0$, $S(p) = 1$ and $C(x) > 0$ for all $x \in [0, p)$.

Hint: $C(x)$ can be written $1 - \frac{x^2}{2} + \frac{x^4}{24} - \sum_{k=1}^{\infty}\left(\frac{x^{4k+2}}{(4k+2)!} - \frac{x^{4k+4}}{(4k+4)!}\right)$.

(v) S, C are periodic on $\mathbb{R}$ with period $4p$. Hint: Start by defining $\widetilde{C}(x) = S(x + p)$ and $\widetilde{S}(x) = -C(x + p)$ and use the uniqueness result of (iii) above.

Note: The number $2p$, i.e., twice the value of the first positive zero of $\cos x$, is rather important, and we have a special name for it—it usually denoted by π. Calculation shows that $\pi = 3.14159\ldots$.

(vi) $\gamma(x) = (C(x), S(x))$, $x \in [0, 2\pi]$ is a C^1 curve, the mapping $\gamma|[0, 2\pi)$ is a 1:1 map of $[0, 2\pi)$ onto the unit circle S^1 of $\mathbb{R}^2$, and the arc-length $S(t)$ of the part of the curve $\gamma|[0, t]$ is t for each $t \in [0, 2\pi]$. (See Figure A.2.)

Remark: Thus, we can geometrically think of the angle between e_1 and $(C(t), S(t))$ as t (that's t radians, meaning the arc on the unit circle going from e_1 to $P = (C(t), S(t))$ (counterclockwise) has length t as you are asked to prove in the above question) and we have the geometric interpretation that $C(t)(= \cos t)$ and $S(t)(= \sin t)$ are, respectively, the lengths of the adjacent and the opposite sides of the right triangle with vertices at $(0, 0)$, $(0, C(t))$, $(C(t), S(t))$ and angle t at the vertex $(0, 0)$, at least if $0 < t < p = \frac{\pi}{2}$. (Notice this is now a theorem concerning $\cos t$, $\sin t$ as distinct from a definition.)

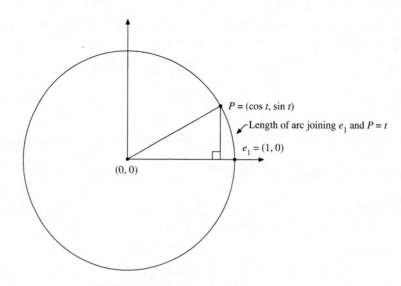

Figure A.2:

Note: Part of the conclusion of (vi) is that the length of the unit circle is 2π. (Again, this becomes a theorem; it is not a definition—π is *defined* to be $2p$, where p, as in (iv), is the first positive zero of the function $\cos x = \sum_{n=0}^{\infty}(-1)^n \frac{x^{2n}}{(2n)!}$.)

LECTURE 7: COMPLEX SERIES, PRODUCTS OF SERIES, AND COMPLEX EXPONENTIAL SERIES

In Real Analysis Lecture 4 we discussed series $\sum_{n=1}^{\infty} a_n$ where $a_n \in \mathbb{R}$. One can analogously consider complex series, i.e., the case when $a_n \in \mathbb{C}$. The definition of convergence is exactly the same as in the real case. That is we say the series converges if the nth partial sum (i.e., $\sum_{j=1}^{n} a_j$) converges; more precisely:

Definition: $\sum_{n=1}^{\infty} a_n$ converges if the sequence of partial sums $\{\sum_{j=1}^{n} a_j\}_{n=0,1,\ldots}$ is a convergent sequence in $\mathbb{C}$; that is, there is a complex number $s = u + iv$ (u, v real) such that $\lim_{n\to\infty} \sum_{j=1}^{n} a_j = s$.

Note: Of course here we are using the terminology that a sequence $\{z_n\}_{n=1,2,\ldots} \subset \mathbb{C}$ converges, with limit $a = \alpha + i\beta$, if the real sequence $|z_n - a|$ has limit zero, i.e., $\lim_{n\to\infty} |z_n - a| = 0$. In terms of "$\varepsilon, N$" this is the same as saying that for each $\varepsilon > 0$ there is N such that $|z_n - a| < \varepsilon$ for all $n \geq N$. Writing z_n in terms of its real and imaginary parts, $z_n = u_n + iv_n$, then this is the same as saying $u_n \to \alpha$ and $v_n \to \beta$, so applying this to the sequence of partial sums we see that the complex series $\sum_{n=1}^{\infty} a_n$ with $a_n = \alpha_n + i\beta_n$ is convergent if and only if both of the real series $\sum_{n=1}^{\infty} \alpha_n$, $\sum_{n=1}^{\infty} \beta_n$ are convergent, and in this case $\sum_{n=1}^{\infty} a_n = \sum_{n=1}^{\infty} \alpha_n + i \sum_{n=1}^{\infty} \beta_n$.

Most of the theorems we proved for real series carry over, with basically the same proofs, to the complex case. For example, if $\sum_{n=1}^{\infty} a_n$, $\sum_{n=1}^{\infty} b_n$ are convergent complex series then the series $\sum_{n=1}^{\infty} (a_n + b_n)$ is convergent and its sum (i.e., $\lim_{n\to\infty} \sum_{j=1}^{n}(a_j + b_j)$) is just $\sum_{n=1}^{\infty} a_n + \sum_{n=1}^{\infty} b_n$.

Also, again analogously to the real case, we say the complex series $\sum_{n=1}^{\infty} a_n$ is absolutely convergent if $\sum_{n=1}^{\infty} |a_n|$ is convergent. We claim that just as in the real case absolute convergence implies convergence.

Lemma 1. *The complex series $\sum_{n=1}^{\infty} a_n$ is convergent if $\sum_{n=1}^{\infty} |a_n|$ is convergent.*

Proof: Let α_n, β_n denote the real and imaginary parts of a_n, so that $a_n = \alpha_n + i\beta_n$ and $|a_n| = \sqrt{\alpha_n^2 + \beta_n^2} \geq \max\{|\alpha_n|, |\beta_n|\}$, so $\sum_{n=1}^{\infty} |a_n|$ converges $\Rightarrow \exists$ a fixed $M > 0$ such that $\sum_{j=1}^{n} |a_j| \leq M$ $\forall n \Rightarrow \sum_{j=1}^{n} |\alpha_j| \leq M$ $\forall n \Rightarrow \sum_{n=1}^{\infty} |\alpha_n|$ is convergent, so $\sum_{n=1}^{\infty} \alpha_n$ is absolutely convergent, hence convergent. Similarly, $\sum_{n=1}^{\infty} \beta_n$ is convergent. But $\sum_{j=1}^{n} a_j = \sum_{j=1}^{n} \alpha_j + i \sum_{j=1}^{n} \beta_j$ and so $\lim_{n\to\infty} \sum_{j=1}^{n} a_j$ exists and equals $\lim_{n\to\infty} \sum_{j=1}^{n} \alpha_j + i \lim_{n\to\infty} \sum_{j=1}^{n} \beta_j = \sum_{n=1}^{\infty} \alpha_n + i \sum_{n=1}^{\infty} \beta_n$, which completes the proof.

We next want to discuss the important process of multiplying two series: If $\sum_{n=0}^{\infty} a_n$ and $\sum_{n=0}^{\infty} b_n$ are given complex series, we observe that the product of the partial sums, i.e., the product $(\sum_{n=0}^{N} a_n) \cdot (\sum_{n=0}^{N} b_n)$, is just the sum of all the elements in the rectangular array

$$
\begin{pmatrix}
a_N b_0 & a_N b_1 & a_N b_2 & \cdots & \cdots & \cdots & a_N b_N \\
a_{N-1} b_0 & a_{N-1} b_1 & a_{N-1} b_2 & \cdots & \cdots & \cdots & a_{N-1} b_N \\
\vdots & \vdots & \vdots & & & & \vdots \\
\vdots & \vdots & \vdots & & & & \vdots \\
a_2 b_0 & a_2 b_1 & a_2 b_2 & \cdots & \cdots & \cdots & a_2 b_N \\
a_1 b_0 & a_1 b_1 & a_1 b_2 & \cdots & \cdots & \cdots & a_1 b_N \\
a_0 b_0 & a_0 b_1 & a_0 b_2 & \cdots & \cdots & \cdots & a_0 b_N
\end{pmatrix}
$$

$$
\left(\sum_{n=0}^{N} a_n\right)\left(\sum_{n=0}^{N} b_n\right) = \sum_{n=0}^{2N}\left(\sum_{0\le i,j\le N,\, i+j=n} a_i b_j\right),
$$

and observe that if $i, j \ge 0$ and $i + j \le N$ we automatically have $i, j \le N$, and so with $c_n = \sum_{i,j\ge 0,\, i+j=n} a_i b_j \ (= \sum_{i=0}^{n} a_i b_{n-i})$ we see that the right side of the above identity can be written $\sum_{n=0}^{N} c_n + \sum_{0\le i,j\le N,\, i+j>N} a_i b_j$, and so we have the identity

$$(*) \qquad \left(\sum_{n=0}^{N} a_n\right)\left(\sum_{n=0}^{N} b_n\right) - \sum_{n=0}^{N} c_n = \sum_{0\le i,j\le N,\, i+j>N} a_i b_j$$

for each $N = 0, 1, \ldots$ (Geometrically, $\sum_{n=0}^{N} c_n$ is the sum of the lower triangular elements of the array, including the leading diagonal, and the term on the right of $(*)$ is the sum of the remaining, upper triangular, elements.) If the given series $\sum a_n, \sum b_n$ are absolutely convergent, we show below that the right side of $(*) \to 0$ as $N \to \infty$, so that $\sum_{n=0}^{\infty} c_n$ converges and has sum equal to $\left(\sum_{n=0}^{\infty} a_n\right)\left(\sum_{n=0}^{\infty} b_n\right)$. That is:

Lemma (Product Theorem.) *If $\sum a_n$ and $\sum b_n$ are absolutely convergent complex series, then $\left(\sum_{n=0}^{\infty} a_n\right)\left(\sum_{n=0}^{\infty} b_n\right) = \sum_{n=0}^{\infty} c_n$, where $c_n = \sum_{i=0}^{n} a_i b_{n-i}$ for each $n = 0, 1, 2, \ldots$; furthermore, the series $\sum c_n$ is absolutely convergent.*

Proof: By $(*)$ we have

$$
\left|\left(\sum_{i=0}^{N} a_i\right)\left(\sum_{j=0}^{N} b_j\right) - \sum_{n=0}^{N} c_n\right| = \left|\sum_{i,j\le N,\, i+j>N} a_i b_j\right| \le \sum_{i,j\le N,\, i+j>N} |a_i||b_j|
$$

$$
\le \sum_{i,j\le N,\, i>N/2} |a_i||b_j| + \sum_{i,j\le N,\, j>N/2} |a_i||b_j|
$$

$$
= \left(\sum_{i=[N/2]+1}^{N} |a_i|\right)\left(\sum_{j=0}^{N} |b_j|\right)\left(\sum_{i=0}^{N} |a_i|\right)\left(\sum_{j=[N/2]+1}^{N} |b_j|\right)
$$

$$\leq \Big(\sum_{i=[N/2]+1}^{\infty} |a_i| \Big) \Big(\sum_{j=0}^{\infty} |b_j| \Big) + \Big(\sum_{i=0}^{\infty} |a_i| \Big) \Big(\sum_{j=[N/2]+1}^{\infty} |b_j| \Big)$$

$$\to 0 \text{ as } N \to \infty,$$

where $[N/2] = \frac{N}{2}$ if N is even and $[N/2] = \frac{N-1}{2}$ if N is odd. Notice that in the last line we used the fact that $\sum_{i=[N/2]+1}^{\infty} |a_i| = \sum_{i=0}^{\infty} |a_i| - \sum_{i=0}^{[N/2]} |a_i| \to 0$ as $N \to \infty$ because by definition $\sum_{n=0}^{\infty} |a_n| = \lim_{J \to \infty} \sum_{n=0}^{J} |a_n|$, and similarly, $\sum_{j=[N/2]+1}^{\infty} |b_j| = \sum_{j=0}^{\infty} |b_j| - \sum_{j=0}^{[N/2]} |b_j| \to 0$, because $\sum_{n=0}^{\infty} |b_n| = \lim_{J \to \infty} \sum_{n=0}^{J} |b_n|$.

This completes the proof that $\sum_{n=0}^{\infty} c_n$ converges, and $\sum_{n=0}^{\infty} c_n = \left(\sum_{n=0}^{\infty} a_n \right) \left(\sum_{n=0}^{\infty} b_n \right)$.

To prove that $\sum_{n=0}^{\infty} |c_n|$ converges, just note that for each $n \geq 0$ we have $|c_n| \equiv \left| \sum_{i+j=n} a_i b_j \right| \leq \sum_{i+j=n} |a_i||b_j| \equiv C_n$ say, and the above argument, with $|a_i|, |b_j|$ in place of a_i, b_j, respectively, shows that $\sum_{n=0}^{\infty} C_n$ converges, so by the comparison test $\sum_{n=0}^{\infty} |c_n|$ also converges.

We now define the complex exponential series.

Definition: The complex exponential function $\exp z$ (also denoted e^z) is defined by $\exp z = \sum_{n=0}^{\infty} \frac{z^n}{n!}$.

Observe that this makes sense for all z because the series $\sum_{n=0}^{\infty} \frac{|z|^n}{n!}$ is the real exponential series, which we know is convergent on all of $\mathbb{R}$, so that the series $\sum_{n=0}^{\infty} \frac{z^n}{n!}$ is absolutely convergent (hence convergent by Lem. 1) for all z.

We can use the above product theorem to check the following facts, which explains why the notation e^z is sometimes used instead of $\exp z$:

(i) $\qquad\qquad\qquad (\exp a)(\exp (b) = \exp(a + b), \qquad a, b \in \mathbb{C},$

(ii) $\qquad\qquad\qquad \exp ix = \cos x + i \sin x, \qquad x \in \mathbb{R}.$

The proof is left as an exercise (Exercises 7.1 and 7.2 below).

Notice that it follows from (i) that $\exp z$ is never zero (because by (i) $(\exp z)(\exp -z) = \exp 0 = 1 \neq 0 \; \forall z \in \mathbb{C}$).

LECTURE 7 PROBLEMS

7.1 Use the product theorem to show that $\exp a \; \exp b = \exp(a + b)$ for all $a, b \in \mathbb{C}$.

7.2 Justify the formula $\exp ix = \cos x + i \sin x$ for all $x \in \mathbb{R}$.

Note: cos, sin are defined by $\cos x = \sum_{k=0}^{\infty} (-1)^k \frac{x^{2k}}{(2k)!}$ and $\sin x = \sum_{k=0}^{\infty} (-1)^k \frac{x^{2k+1}}{(2k+1)!}$ for all real x.

LECTURE 8: FOURIER SERIES

So far we have only considered vectors in $\mathbb{R}^n$, but of course the concept of vector is really much more general. Indeed, any collection of objects V which have a well defined operation of addition and multiplication by real scalars such that following eight *vector space axioms* are satisfied, is called a *real vector space* and the objects themselves are called *vectors*:

8.1 Vector Space Axioms for V:

We assume that for $v, w \in V$ and $\lambda \in \mathbb{R}$ there are well-defined operations "+" (addition) and "." (multiplication of vector by a scalar) to give $v + w \in V$ and $\lambda \cdot v \in V$ (henceforth written just as λv) and that the following properties hold:

(i) $v, w \in V \Rightarrow v + w = w + v$

(ii) $u, v, w \in V \Rightarrow (u + v) + w = u + (v + w)$

(iii) $\exists 0 \in V$ (called the "zero vector") with $v + 0 = v \; \forall v \in V$

(iv) $v \in V \Rightarrow \exists$ an element $-v$ with $v + (-v) = 0$

(v) $\lambda, \mu \in \mathbb{R}, \; v \in V \Rightarrow \lambda(\mu v) = (\lambda \mu) v$

(vi) $\lambda, \mu \in \mathbb{R}, \; v \in V \Rightarrow (\lambda + \mu) v = (\lambda v) + (\mu v)$

(vii) $\lambda \in \mathbb{R}, \; v, w \in V \Rightarrow \lambda(v + w) = (\lambda v) + (\lambda w)$

(viii) $v \in V \Rightarrow 1 v = v$.

Note: In a general vector space we usually write $v - w$ as an abbreviation for $v + (-w)$.

There are many examples of such general real vector spaces: For example, the set $C^0([a, b])$ of continuous functions $f : [a, b] \to \mathbb{R}$ trivially satisfies the above axioms if addition and multiplication by scalars are defined pointwise (i.e., $f, g \in C^0([a, b]) \Rightarrow f + g \in C^0([a, b])$ is defined by $(f + g)(x) = f(x) + g(x) \forall x \in [a, b]$, and $\lambda \in \mathbb{R}$ and $f \in C^0([a, b]) \Rightarrow \lambda f$ is defined by $(\lambda f)(x) = \lambda f(x)$ for $x \in [a, b]$). In this case, the zero vector is the zero function $0(x) = 0 \forall x \in [a, b]$. Most of the definitions and theorems for the Euclidean space $\mathbb{R}^n$ carry over with only notational changes to general real vector spaces. For example, the linear dependence lemma holds in a general vector space and the basis theorem applies in any subspace which is the span of finitely many vectors.

Suppose now that V is a real vector space equipped with a *inner product* $\langle \, , \, \rangle$. Thus, $\langle v, w \rangle$ is real valued and

(i) $\langle v, w \rangle = \langle w, v \rangle \quad \forall v, w \in V$,

(ii) $\langle \alpha v_1 + \beta v_2, w \rangle = \alpha \langle v_1, w \rangle + \beta \langle v_2, w \rangle \quad \forall v_1, v_2, w \in V$ and $\alpha, \beta \in \mathbb{R}$, and

(iii) $\langle v, v \rangle \geq 0 \quad \forall v \in V$ with equality if and only if $v = 0$.

We define the *norm* of V, denoted $\| \; \|$, by $\|v\| = \sqrt{\langle v, v \rangle}$.

Notice that then the inner product and the norm have properties very similar to the dot product and norm of vectors in $\mathbb{R}^n$ (indeed an important special case of inner product is the case $V = \mathbb{R}^n$ and $\langle v, w \rangle = v \cdot w$). In the general case, by properties (i), (ii), (iii), we have

(1) $$\|v - w\|^2 = \|v\|^2 + \|w\|^2 - 2\langle v, w \rangle \quad \forall \, v, w \in V \,,$$

and we can prove various other results which have analogues in the special case when $V = \mathbb{R}^n$ and $\langle v, w \rangle = v \cdot w$; for example, there is a Cauchy-Schwarz inequality (proved exactly as in $\mathbb{R}^n$):

$$|\langle v, w \rangle| \leq \|v\| \, \|w\|, \quad v, w \in V \,.$$

Analogous to the $\mathbb{R}^n$ case, a set $v_1, \ldots, v_N$ of vectors in V is said to be *orthonormal* if

(2) $$\langle v_i, v_j \rangle = \delta_{ij} \quad \forall \, i, j \in \{1, \ldots, N\}$$

where δ_{ij} denotes the "Kronecker delta," defined by

$$\delta_{ij} = \begin{cases} 1 & \text{if } i = j \\ 0 & \text{if } i \neq j \,. \end{cases}$$

Given such an orthonormal set $v_1, \ldots, v_N$ and given any other $v \in V$, we define

(3) $$c_n = \langle v, v_n \rangle \,.$$

Then by using (1) we get

$$\|v - \textstyle\sum_{n=1}^{N} c_n v_n\|^2 = \|v\|^2 + \textstyle\sum_{n=1}^{N} \sum_{m=1}^{N} c_n c_m \langle v_n, v_m \rangle - 2 c_n \langle v, v_n \rangle \,.$$

And hence by (2), (3) we get

(*) $$\|v - \textstyle\sum_{n=1}^{N} c_n v_n\|^2 = \|v\|^2 - \textstyle\sum_{n=1}^{N} c_n^2 \,,$$

and hence in particular,

(**) $$\textstyle\sum_{n=1}^{N} c_n^2 \leq \|v\|^2 \,,$$

with equality if and only if $v = \sum_{n=1}^{N} c_n v_n$. (Notice that geometrically $\sum_{n=1}^{N} c_n v_n$ represents the "orthogonal projection" of v onto the subspace spanned by $v_1, \ldots v_N$—see Exercise 8.1.)

Now in the case when V is not finite dimensional we can have an orthonormal sequence $v_1, v_2 \ldots$ of elements of V. In particular, the above discussion applies to $v_1, v_2, \ldots, v_N$ for each integer $N \geq 1$, and (**) tells us that the partial sums $\sum_{n=1}^{N} c_n^2$ of the series $\sum_{n=1}^{\infty} c_n^2$ are bounded above by $\|v\|^2$. Hence, $\sum_{n=1}^{\infty} c_n^2$ converges (recall from Lecture 4 that a series of nonnegative terms converges if and only if the partial sums are bounded) and

(‡) $$\sum_{n=1}^{\infty} c_n^2 \leq \|v\|^2 \,.$$

($\ddagger$) is called *Bessel's Inequality*. Notice (by ($*$)) that *equality holds in* ($\ddagger$) *if and only if* $\lim_{N\to\infty} \|v - \sum_{n=1}^{N} c_n v_n\| = 0$. This ties in with the following definition.

Definition: The orthonormal sequence $v_1, v_2, \ldots$ is said to be *complete* if, for each $v \in V$, we have $\lim_{N\to\infty} \|v - \sum_{n=1}^{N} c_n v_n\| = 0$; here of course $c_n = \langle v, v_n \rangle$ as in (3) above. Notice that by ($*$) *completeness holds if and only if equality holds in* ($\ddagger$) $\forall v \in V$.

Terminology: If $\lim_{N\to\infty} \|v - \sum_{n=1}^{N} c_n v_n\| = 0$ then we say the series $\sum_{n=1}^{\infty} c_n v_n$ *converges to* v *in the norm* $\| \|$ of V. Whether or not this convergence takes place, the series $\sum_{n=1}^{\infty} c_n v_n$ is called *the Fourier series for* v *relative to the orthonormal sequence* $v_1, v_2, \ldots$.

For the rest of this lecture we specialize to a concrete situation—Viz., Fourier series for piecewise continuous functions. First we introduce some notations: for any given closed interval $[a, b] \subset \mathbb{R}$, $C_P([a, b]) =$ the set of *piecewise continuous* functions on $[a, b]$. That is $f \in C_P([a, b])$ means there is a partition $a = x_0 < x_1 < \ldots < x_J = b$ of the interval $[a, b]$ such that f is continuous at each point of (x_{j-1}, x_j), $j = 1, \ldots, J$, and such that

$$\lim_{x\downarrow a} f(x), \ \lim_{x\uparrow b} f(x), \ \lim_{x\downarrow x_j} f(x), \ \lim_{x\uparrow x_j} f(x) \text{ all exist } \quad j = 1, \ldots, J - 1 .$$

$C_A([a, b])$ denotes the subset of $C_P([a, b])$ functions f with the additional properties that

$$f(a) = f(b) = \frac{1}{2}\left(\lim_{x\downarrow a} f(x) + \lim_{x\uparrow b} f(x)\right)$$

and

$$f(x_j) = \frac{1}{2}\left(\lim_{x\downarrow x_j} f(x) + \lim_{x\uparrow x_j} f(x)\right), \quad j = 1, \ldots, J - 1 .$$

One readily checks that if $f, g \in C_P([a, b])$ and if $\alpha, \beta \in \mathbb{R}$, then $\alpha f + \beta g \in C_P([a, b])$, so $C_P([a, b])$ is a vector space which is a subspace of the space of all functions $f : [a, b] \to \mathbb{R}$. Likewise, $C_A([a, b])$ is a vector space, which is a subspace of $C_P([a, b])$.

We now specialize further to the interval $[-\pi, \pi]$ and we define an inner product on $C_A([a, b])$ by

($\dagger$) $$\langle f, g \rangle = \frac{1}{\pi} \int_{-\pi}^{\pi} f(x)g(x)\,dx ,$$

so that $\|f\| = \sqrt{\frac{1}{\pi} \int_{-\pi}^{\pi} f^2(x)\,dx}$. (The factor $\frac{1}{\pi}$ is present purely for technical reasons.)

By direct computation, one checks that the sequence $\frac{1}{\sqrt{2}}, \cos x, \sin x, \cos 2x, \sin 2x, \ldots,$ $\cos nx, \sin nx, \ldots$ is an orthonormal sequence relative to this inner product (Exercise 8.4.).

Definition: Given $f \in C_A([-\pi, \pi])$, the trigonometric Fourier series of f is defined to be the Fourier series of f relative to the orthonormal sequence $\frac{1}{\sqrt{2}}, \cos x, \sin x, \cos 2x, \sin 2x,$ $\ldots, \cos nx, \sin nx, \ldots$. Thus, the trigonometric Fourier series of f can be written

$$c_0 \frac{1}{\sqrt{2}} + \sum_{n=1}^{\infty} (a_n \cos nx + b_n \sin nx),$$

where